THE NATURE OF TECHNOLOGICAL KNOWLEDGE. ARE MODELS OF SCIENTIFIC CHANGE RELEVANT?

SOCIOLOGY OF THE SCIENCES
MONOGRAPHS

1984

THE NATURE OF TECHNOLOGICAL KNOWLEDGE. ARE MODELS OF SCIENTIFIC CHANGE RELEVANT?

Edited by

RACHEL LAUDAN
Center for the Study of Science in Society,
Virginia Polytechnic Institute and State University, Blacksburg, U.S.A.

D. REIDEL PUBLISHING COMPANY
A MEMBER OF THE KLUWER ACADEMIC PUBLISHERS GROUP
DORDRECHT / BOSTON / LANCASTER

Library of Congress Cataloging in Publication Data

Main entry under title:

The Nature of Technological Knowledge.

(Sociology of the sciences monographs)
Includes index.
1. Technology–Philosophy–Congresses. 2. Technological innovations–Congresses. 3. Science–Philosophy–Congresses.
I. Laudan, Rachel, 1944– II. Series.
T14.N37 1984 306'.46 83-24642
ISBN 90-277-1716-8

Published by D. Reidel Publishing Company,
P.O. Box 17, 3300 AA Dordrecht, Holland.

Sold and distributed in the U.S.A. and Canada
by Kluwer Academic Publishers,
190 Old Derby Street, Hingham, MA 02043, U.S.A.

In all other countries, sold and distributed
by Kluwer Academic Publishers Group,
P.O. Box 322, 3300 AH Dordrecht, Holland.

Printed in The Netherlands

CONTENTS

NOTES ON CONTRIBUTORS

EDWARD CONSTANT is Associate Professor in the Department of History and Philosophy at Carnegie-Mellon University. He is author of *Origins of the Turbojet Revolution*, and is currently working on a study of the oil industry.

GARY GUTTING is Professor of Philosophy at Notre Dame University. He is author of *Paradigms and Revolutions: Applications and Appraisals of Thomas Kuhn's Philosophy of Science* and has published extensively on the philosophy of science and on continental philosophy.

NORMAN HUMMON is Associate Professor in the Department of Sociology at the University of Pittsburgh. He received a B.S. in Electrical Engineering from the University of Michigan, an M.A. in Administration Sciences from Yale University and his Ph.D. from Cornell University. His research interests concentrate on the sociology of technology, and on quantitative methods.

RACHEL LAUDAN is Associate Professor of Science and Technology Studies at Virginia Polytechnic Institute and State University. She received her Ph.D. in History and Philosophy of Science from the University of London, and has published a number of articles on the history of the earth sciences. Her research interests are in the history of geology and the history of technology.

DEREK J. DESOLLA PRICE was Avalon Professor in the Department of History of Science and Medicine at Yale University. He was author of *Little Science, Big Science* and *Science Since Babylon* as well as numerous articles and books on the history of technology.

PETER WEINGART, born 1941, is professor of sociology of science and science policy at the University of Bielefeld since 1973; author of *Die amerikanische Wissenschaftslobby* (1970) *Wissensproduktion und soziale Struktur* (1976); author, co-author and editor of numerous books and articles in the area of sociology of science, technology and science policy studies. In 1983 he held a visiting appointment at the Institute for Advanced Studies in Berlin.

Rachel Laudan

INTRODUCTION

One of the ironies of our time is the sparsity of useful analytic tools for understanding change and development within technology itself. For all the diatribes about the disastrous effects of technology on modern life, for all the equally uncritical paeans to technology as the panacea for human ills, the vociferous pro- and anti-technology movements have failed to illuminate the nature of technology. On a more scholarly level, in the midst of claims by Marxists and non-Marxists alike about the technological underpinnings of the major social and economic changes of the last couple of centuries, and despite advice given to government and industry about managing science and technology by a small army of consultants and policy analysts, technology itself remains locked inside an impenetrable black box, a *deus ex machina* to be invoked when all other explanations of puzzling social and economic phenomena fail.

The discipline that has probably done most to penetrate that black box in recent years by studying the internal development of technology is history.[1] Historians of technology and certain economic historians have carried out careful and detailed studies on the genesis and impact of technological innovations, and the structure of the social systems associated with those innovations. Within the past few decades tentative consensus about the periodization and the major traditions within the history of technology has begun to emerge, at least as far as Britain and America in the eighteenth and nineteenth century are concerned. Most historians would now agree about the outlines of technological history, and about the areas where further research is most needed. While younger scholars will doubtless produce new interpretations, they will do so against the backdrop of the coherent narrative constructed by the pioneering generations of historians.[2]

But if a preliminary narrative account of the recent history of western technology is now available, it remains the case that we are still in want of some theoretical

R. Laudan (ed.), The Nature of Technological Knowledge, 1–26.

generalizations about the causes and mechanisms of technological change. The preparatory stage of providing empirical historical data to guide the construction of models may have been accomplished, but model-building itself remains embryonic. Even many of the general conclusions that are implicitly accepted as truisms by historians of technology are largely unknown and unappreciated outside that very small community. Yet attempts to construct models of technological change without a firm grounding in recent work in the history of technology are likely to prove inadequate. Acutely aware of this problem, many of the leading historians of technology are venturing well beyond their normal aversion to generalizations, and advancing general principles, and even models. These serve to draw together published historical work, to guide further research, and should be of interest to the many disciplines outside history that have a stake in understanding technological change.[3]

Of course, the construction of models necessarily over-simplifies the rich texture of the past, and fails to draw attention to the complex concatenation of themes interacting in any episode. Provided this is recognized, it need not give us pause since the purpose of developing a model is not, after all, to reproduce history, but rather to simplify and to throw into relief those aspects of the historical record of most interest to the investigator. A model that incorporated every aspect of the system it was designed to model would fail to serve this function, being no more than a re-creation of the original system. In order to understand how technological change occurs a sustained effort to construct tentative and corrigible general theories is necessary.

Technological change, as generally understood, is a multi-faceted phenomenon involving cognitive, social, organizational and economic factors to name but a few. It has been studied by scholars from disciplines as diverse as economics and anthropology, business management and history, political science and sociology. Rather than tackle technological change *tout court*, it seems more reasonable at the present time to divide the problem, and hope to conquer it by this means. Besides the well-known economic, political and social influence on technological change, shifts in the knowledge of the practitioners play a crucial role in technological development. This volume concentrates on understanding those cognitive aspects of technological change. Two further assumptions are made.

First, that since the most thoroughly studied form of knowledge change is science, it is appropriate to explore analogies and disanalogies between changing scientific knowledge and changing technological knowledge. Second, that an informed knowledge of the history of technology constitutes an important basis for understanding technological change.

Before these assumptions are explored in more depth, two potential criticisms need to be addressed. First, why should it be assumed that the most appropriate data for models of cognitive technological change come from history rather than from sociological and economic surveys of contemporary technology; and second, why, develop alternatives since there are already plenty of economic models for understanding technological change? Ready responses can be made to both criticisms. First the sociological data base for contemporary technology is, in fact, meager in the extreme, sociologists having shown scant interest in technology.[4] Economic data are much more plentiful, indeed so plentiful that definitive conclusions are hard to draw. But even if appropriate sociological and economic evidence were already available, history still has an important role to play when we seek to understand the long term as well as short term patterns in the development of technology. Furthermore the historical record gives an invaluable insight into the cognitive development of technology.

As to the second criticism, no one who looks at the diverse literature on technological change would deny that economic models make up the bulk of the theoretically-oriented discussions of technology.[5] Yet economic analyses (together with cognate organizational and social analyses) typically fail to shed light on the *internal* dynamics of technological change. Rather, economists seek to determine the relationship between inventions and innovations on the one hand and economic changes, usually productivity, on the other. Whether of a 'demand-pull' or 'technology-push' type, the models that are produced have too gross a structure to capture the internal dynamics of technological change.[6]

Nor should the economists' failure to analyze the internal development of technology be surprising, since their interest is in the effect of technology on the economy rather than in technology *per se*. For many economists the spur to the study of technology was the realization that the rapid economic growth of the west in the past hundred years could not be explained solely, or even mainly in traditional economic terms of the increasing stock of capital per worker.[7] Productivity increases as a result of changing technologies were seized on as a plug for the subsequent explanatory gap. In short, the economists' aim was to understand economic change, not technological change.[8] To say that economic analyses, which take technology as an exogenous variable and technological change as a given, do little to contribute to our understanding of the internal development of technology is no criticism of their importance or validity. It is simply to recognize that studies that treat technology in this way are likely to be of only peripheral interest to those for whom technology itself is the central focus of concern.

In sum, what is being called for here is an account, model or theory of the dynamics of cognitive change in technology. Many question whether this is possible, or even if possible, whether it is worthwhile given the apparently overwhelming importance of economic and social factors in stimulating and selecting new technologies. I would argue that it is not only possible but essential. External factors may in many cases be responsible for the problems the technological practitioner chooses to study. Equally external factors may determine which technologies are eventually adopted. Nonetheless the problems have to be perceived as soluble by technological means, and the technologies themselves have to be created. Without such internal technological capabilities all the economic and social pressures in the world would be to no avail. To believe that technological change can be explained entirely in external terms is to have a very optimistic view of technology's ability deliver. It is to assume that given any external stimulus the mechanician or engineer will always be able to develop an appropriate technology in response. Yet historical work shows us that this is patently not the case.[9] Technology has its own internal dynamic, and unless we understand this we shall be incapable of understanding how technologists respond to social and economic pressures, or how their work subsequently affects society at large.

Thus I want to distinguish the different meanings of the expression 'technological change'. It has become such an omnibus term, referring variously to the social and economic roots of such change, the details of the research and development cycle, the effects of technology on industrial, social and military life, and even to technology 'out of control', that it is easy to talk at cross purposes.[10] The problem is compounded by the fact that there is as much diversity in the definition of technology as there is in the understanding of technological change. Ranging from the continental tradition with its distinction between technology and technique, through the amateur historian's fixation on hardware, to ambitious recent attempts to include all institutionalized social processes under the umbrella of technology, the concept staggers under the interpretive load it has to carry. Since no one scholar can work with all these concepts of technology simultaneously, and since the hope of all these facets of technological change at once seems remote, the issue of the notion of technology and technological change adopted in this anthology has to be addressed.

With respect to the definition of technology, the commonsense approach of accepting certain artifacts and processes (those normally studied under the rubric of the history of technology) as exemplary has been adopted. Rather than spend time in probably fruitless attempts to demarcate technology from other activities, it seems preferable to work from a set of cases – steam engines and railroads, turbojets and chemical plants – that seem intuitively paradigmatic. As for technological change, focusing on the skilled and purposeful actions of a handful of practitioners, applying their accumulated knowledge to solve technological problems, seems a fruitful approach. As these practitioners develop new technologies, the pool of knowledge that they share shifts. Thus an important avenue to understanding technological change is to analyze how the technological knowledge shared by a community of practitioners shifts. The subjects of study are the knowledge-generating and diffusing activities of technological practitioners.

Viewed in this light, the models of technological change become a special case of knowledge change. Analyses that are already in hand for understanding patterns of knowledge generation and acceptance ought to be potentially applicable to technology. One genre of knowledge change that has been intensively studied in recent decades is scientific change. Stimulated by the publication of Thomas Kuhn's influential *Structure of Scientific Revolutions* in 1962, historians, philosophers and sociologists of science have turned to problem of the development of scientific knowledge with renewed vigor. An extensive body of literature exists dealing with the historical development, philosophical significance and accompanying social structures of scientific knowledge. The extent to which categories developed for scientific change are applicable to technology deserves fuller examination.

Given the popularity of applying Kuhnian analyses to almost every human activity, this might seem an unproblematic program.[11] Yet those in the business of technology studies have been very leary of treating technological change as knowledge change, and particularly so if it were thought of as analogous to scientific change. Indeed technology is almost unique among disciplines in having been the subject of only the occasional Kuhnian analysis. There are, I believe, three reasons why this has been so: first, the assumption that technological knowledge is quintessentially tacit; second, the identification of technological knowledge with applied science; and third, the selection of analytical units for the history and present structure of technology that, however useful for some purposes, do little to throw the cognitive aspect of technology into prominence. During the past decade or so, a number of different developments in the study of technology have made these barriers seem less formidable than previously. Concomitantly, the changes in history and philosophy of science mentioned above have made its application to technology seem much less threatening.

The widely-held, if often unspoken, assumption that technological knowledge is largely inaccessible to scholarly study seems to be based on the following reasoning: since technological knowledge is rarely articulated, and since when articulated, such knowledge is largely in visual, rather than verbal or mathematical form it does not lend itself to analysis by a scholarly community trained primarily in the analysis of texts and the explication of logical structures. Technological knowledge, on this construal, is 'tacit' knowledge.[12] Technological activities cannot be fully specified, and hence rules for their performance cannot be spelled out. Transmission of such knowledge is therefore limited to the range of personal contacts. Although the term 'tacit knowledge' is not generally current among students of technology, the concept certainly is, supported by two widely-cited pieces of evidence.

The first comes from the realm of technology transfer. Numerous studies, both of an historical nature (the transfer of British technology to North America for example) and of contemporary cases (the transfer of technology to the developing world) make it transparently clear that technology transfer is very hard to achieve.[13] A substantial part of the difficulty lies in the tacit nature of technological knowledge.[14] The fastest way to transfer technology is for workmen already familiar with the technology to set it up in the new situation, teaching new practitioners as they do so. To send machines and operating manuals without accompanying personnel is to face almost certain failure. This particular problem with technology transfer draws attention to the tacit nature of technological knowledge.

The second argument for technology as 'tacit knowledge' is more exclusively historical. Historians are well aware that technological knowledge can easily be lost. The pool of technological knowledge frequently shifts rather than expands with the appearance of new technologies, and is thus not always cumulative. There are losses as well as gains.[15] If practitioners cease using a particular technology the knowledge of how to use it commonly dies with them. The mute presence of the remaining artifacts does not speak for itself, if there is no explicit statement separate from the artifacts.

The conclusion that has been drawn from these two phenomena is that, since technological knowledge is tacit, it is necessarily opaque to the historian, sociologist or philosopher. He may be able to describe technological artifacts, he may be able to place them in chronological order, to describe the biographies of their inventors, or to trace their effects on society, but not to construct the activities by which the practitioners arrived at these innovations. On this view, changing technological knowledge seems an unpromising subject for model-building.

Ironically some support for this conclusion has been offered by the very historians that have done the most to insist that technology should be studied as knowledge. Scholars, like Eugene Ferguson, who have persistently drawn attention to the cognitive dimension of technology have also gone to great pains to stress the *unique* character of technological knowledge,[16] emphasizing that, especially when compared with science, technology is a highly *visual* activity. Technology is thus isolated from, rather than compared to, other forms of cognitive activity. Since we are relatively ill-equipped for the analysis of knowledge expressed in visual form, further discussion proves very difficult.

Despite these obstacles, recent developments have made the problem of the tacit nature of technological knowledge seem a lot less overwhelming to the student of technological change. In part, this is the result of historical research on the communities of practitioners who over the past two hundred years have been primarily responsible for the production of new technological knowledge. Although the record is far from perfect, these practitioners have left behind them a wealth of written records which historians have been able to mine. For technology since the Industrial Revolution there are numerous sources of information about technology over and above the remaining artifacts. Letters,[17] working diagrams,[18] encyclopedia articles,[19] laboratory notebooks,[20] engineering journals and textbooks,[21] government reports,[22] records of technical societies,[23] and the multitudinous documents of industrial communities[24] all constitute rich sources for the historian. In using these resources historians have demonstrated beyond a shadow of a doubt that we need not tell the history of technology solely in terms of a succession of artifacts, and that technological knowledge is not hopelessly tacit. There are in fact numerous written records that can be used to understand the ways in which practitioners develop and diffuse technological knowledge.

Furthermore the tacit component of technology has clearly decreased since the Industrial Revolution. The rise of the engineering profession, and particularly of formal education in engineering testifies to the extent to which knowledge previously handed down by the craft tradition has been made more explicit.[25] Practitioners are in the process of 'learning how to learn', of making prescriptions for technological activity more explicit, and the learning of technique less subject to the direct passage of tacit knowledge from master to pupil.[26] While it may still be true that the quickest way of transferring technical knowledge is the direct contact of master and pupil, and while it is doubtless the case that the 'art of technology' goes well beyond what can be learned in engineering schools, much of technological knowledge is nonetheless explicit.

Finally, even that paradigmatic case of systematic and explicit knowledge – science – contains many tacit elements.[27] But, relativist sociologists of science notwithstanding, the presence of these tacit elements has not impeded a flourishing and successful tradition of studying scientific knowledge and the way it changes.

That certain aspects of any knowledge-generating activity are normally left tacit is scarcely surprising. If this were not so the process would be maddeningly cumbersome.

Turning now to the second barrier to considering technology as knowledge, we need to confront the widely-held belief that technology is applied science, and the corollary that once we understand the discovery and justification of scientific knowledge, nothing remains to be added about technological knowledge. The origin of this notion is not entirely clear. Certainly it is prominent in the rhetoric of the promoters of science from Francis Bacon and the seventeenth century academicians to James Conant and Vannevar Bush. If the wish is father to the deed, then the hope of generations of supporters of science for a technological payoff from (and hence justification for) their scientific research might well in itself have been adequate to generate the myth. Edwin Layton, in an interesting hypothesis, puts its origin much later, and more squarely within the community of historians of science and technology.[28] He argues that scholars like Rupert Hall, while correctly repudiating the Marxist thesis that the Scientific Revolution was no more than the systematization of the knowledge of the craftsman, overreacted when they came to the converse conclusion, namely that science was prior to, and generative of technology. But whatever the origins of the idea that technology was applied science, it has had extraordinary vitality.[29] No less than three special issues of *Technology and Culture* in the recent years, not to mention countless individual articles and books, have been devoted to revisionist examinations of the relationship of science and technology.[30] But partly because of the difficulty of reaching a wide audience, the myth lives on in the writings of certain philosophers and economists, and ubiquitously in the popular imagination. The specter of technology as a subordinate exercise, the tedious and unexciting result of applying the results of science to practical ends is hard to exorcise.

One even still occasionally encounters the claim that technology is a form of science since its practitioners attempt to solve problems rationally and hence apply 'the scientific method'. This trivializes the issue by making the concept of scientific method so wide as to exclude nothing and explain little. Both science and technology are forms of knowledge and at the most general level knowledge can be thought of as generated by a rational problem-solving process. But the differences between forms of knowledge are very significant. Even in the case of science we have painfully come to learn over the past few years that the dream of a universal scientific method was no more than a dream, and that the methods of biologists, geologists and physicists are very different. Similarly we may assume that technology is likely to be distinctively different from science. Exploring disanalogies as well as analogies is essential if we are to understand cognitive change in technology.

Recent attacks on the concept of technology as applied science have employed two strategies, one empirical and one analytic. On the empirical front, historian after historian has chronicled episodes in the development of technology where the major advances owed little or nothing to science. Whether one takes steam power, water power, machine tools, clock making or metallurgy, the conclusion is the same. The technology developed without the assistance of scientific theory, a position summed up by the slogan 'science owes more to the steam engine than the steam engine owes to science'. As a result historians have offered several alternative hypotheses about the science-technology, interaction – technology as illustrative of scientific theory,[31] technology as providing puzzles for scientific theory,[32] and technology as a natural-historical area of study for natural philosophers.[33] More contemporary data point in the same direction. For all their methodological flaws, the government-sponsored surveys, Project Hindsight and TRACES, did demonstrate that the connection between science and technology even in the modern period is much more complex and tenuous than suggested by the popular image of technology as applied science.[34] Indeed the overwhelming consensus among those who have studied the matter is that, except in certain high-technology industries developed since the rise of the electrical and chemical industries in the late nineteenth century, there has been very little technology that can be classified as applied science.

On the analytic front, Edwin Layton's struggle to reintroduce the classic Aristotelian definition of technology as 'systematic knowledge of the useful arts' – a definition which in no way suggested that this knowledge was generated by science and applied by technology – has been helpful at least to historians.[35] While economists, for example, have paid lip service to similar definitions of technology,[36] this definition had made remarkably little difference to their actual research, since, as we have seen, their primary concern is with the economic impact of technology and not with the generation of technological knowledge.

Although Layton's definition of technology as knowledge has gained widespread acceptance in the community of historians of technology, this has not led them (at least until recently) to exploit its parallel but separate relationship with science. As we have seen, scholars like Ferguson (and Layton too) while campaigning for the idea that technology could and should be regarded as knowledge, accompanied this insight by stressing the special nature of technological knowledge. In particular, they have both highlighted the visual, as opposed to the verbal and mathematical nature of technological knowledge. Brooke Hindle pursues the same theme, arguing that artisans generally think differently from scientists and that design is much more important to them.[37] While quite understandable in light of the struggle to see technology as more than simply applied science, this emphasis has blocked any attempt to exploit the possible parallels with science. Indeed for many other historians of technology the fear of their specialty being thought of as simply a subordinate and rather unimportant portion of the history of science is so strong that they have tended to turn to business, social, or economic historians as their natural allies, rather than historians of science. Even where history of technology continues to flourish in history of science departments it has found its most natural allies among social and institutional historians of science. Although the way is now clear to consider technology as knowledge, historians remain understandably reluctant to exploit analogies science. Nonetheless in recent years there have been signs that this traditional aversion to considering technology as knowledge, and hence analogous to science is shifting. Work in Germany and in the United States has raised once again the possibility of integrating insights from the history and philosophy of science into the study of technology.[38]

It is also worth pointing out that, for the purposes of understanding technological change as knowledge change, we do not need to sort out what the relationship between science and technology actually is, and how it has shifted in the past, interesting as that is in its own right. It is sufficient to grant that science and technology are both forms of knowledge, and to explore possible analogies between them. Indeed, in my view, once this task is accomplished we may well be in a better position to understand the relationship between the two activities.

The third impediment to the treatment of technological change as knowledge change has been the very richness and variety of approaches to the study of technology.[39] Picking out which studies are relevant to the concerns outlined here can be very tricky. Some approaches are clearly irrelevant. For example, the study of the widespread cultural and social effects of technology reveals little about the cognitive content of technology. Investigations of this kind, whether by philosophers concerned with the ethical impact of technology,[40] social scientists tracing its diffusion,[41] or historians examining its impact on different societies[42] are all equally tangential to understanding technology as knowledge. Another approach to technology that does little to illuminate its knowledge component is that which concentrates on technology's role in industries and organizations,[43] since one particular industry may employ many different traditions of knowledge, and since different traditions of knowledge may crop up in a variety of industries and organizations. Thus in order to understand cognitive technological traditions we may have to divide up the historical record in an unaccustomed way. Clock and instrument-makers of the late Middle Ages and the Renaissance, for example, produced artifacts with many different social functions, although the skills they were using were the same. Conversely many different industries employed the American system of manufacturing in the nineteenth century, so tracing this particular tradition of knowledge cannot be confined to any one industry (or social use).[44]

Creating more confusion are the many studies that do concentrate on the *internal* development of technology but that pick units of analysis that hinder comparison with recent work in history, philosophy and sociology of science. One major thrust, for example, has been the study of isolated inventions and innovations. Economists have often been concerned to detect the major technological innovations affecting the economy and thus have concentrated on identifying technologies by the magnitude of their economic impacts. Schumpeter, to take a notable case, looked chiefly for major innovations pulled from a continuously available pool of inventions and made economically effective by the efforts of an entrepreneur.[45] The stress on such individual improvements, rather than on traditions of technology, has hindered efforts to see technology as knowledge. Nor was this approach abandoned after the 1930s. Project Hindsight and TRACES, for example, focused on the origins of individual inventions. John Enos' pioneering, and still much quoted, study of innovation in the petroleum industry, in which he assessed the time lag between the invention and economic innovation of a number of devices, has exactly this structure.[46] Classic sociological studies of the origin of invention, rare as they are, rely on the same strategy.[47] Nor are historians exempt from this approach to technology, as histories tracing the invention of various devices have remained commonplace. This view of technology as a series of individual inventions has been as inimical to any sustained analysis of technological knowledge as the view of science as a series of facts and laws discovered in moments of brilliant intuition was to the development of models of science change. Both suggested that, even if considerable thought went into each individual invention or discovery, this in no way constituted a pool of changing and constantly regenerated community-based knowledge.

Nonetheless there *is* an important thread in technology studies, particularly in the history of technology and economic history, that has focused on the internal development of technology in a way that is very relevant to the concept of technology as knowledge. Dominant in the classic work of Usher, very important in the writing of many senior historians of science, and given a somewhat new twist at present, this research has tended to stress the work of the practitioner in his intellectual context.[48] In general scholars with this bent have concentrated on the continuity of technological change, rather than its sharp discontinuities, seeing inventions as improvements on existing technology rather than as individual acts of genius, which facilitates detection of the intellectual traditions on which the practitioner draws. Tracing out the interconnections between different technologies at the same period also draws attention to the increasingly systematic nature of technological knowledge.

In the course of this research historians have realized just how small the community of active generators of new technology has been until very recently.[49] This in turn has led to a new brand of social history of technology, one that studies the social structure that sustains practitioners developing new technologies. As a result we now realize that the problems tackled by the technological community at any given time are generated as much by the current state of technology as by the push and pull of market forces, government demands, or social needs. Knowledge of the weak spots in technological systems generates much technological change and knowledge of technological traditions guides it. Thus the competence of a pool of practitioners is an essential component of the force and direction of technological change which in turn has much more coherence than has often been supposed. Research on traditions of technological knowledge, and on knowledge-generating and preserving technological communities offers rich resources for the student of cognitive technological change. It is no accident that many scholars who have carried out such research have explicitly tried to link their work to general models of knowledge-generation and change.[50]

With the breaching of these three barriers to the study of technology as knowledge, it is now possible for technology to be regarded as analogous to other forms of knowledge. Change and development of technological knowledge can be compared to change and development in different areas. The chance of developing conceptual categories to describe technological change can be enhanced by experimenting with the application of categories already developed elsewhere. Luckily for those interested in such an approach, there have been concomitant developments in the history and philosophy of science that make this a propitious moment to try such an experiment. History and philosophy of science has been particularly active in the study of knowledge change during the last couple of decades. An important if intermittent concern of history and philosophy of science at least since Whewell's massive works of the mid-nineteenth century, the study of the growth and development of science has become a major interest in the last couple of decades. Thomas Kuhn's *Structure of Scientific Revolutions*, has been followed by works by other authors. Toulmin, Lakatos, Laudan and others have developed a series of models intended to capture some of the more important features of scientific change.[51] Despite striking differences between these models, they share a number of assumptions and conclusions. This growing consensus that at least some of the essential features of scientific change have been identified suggests that the time is ripe to see whether some of the analytical categories could also illuminate technology. Traditionally science was thought of as a set of propositions, a collection of theories – a characterization that emphasized its distance and distinctness from technology.[52] More recently, emphasis has been placed on science as an activity, as a process by which scientists come to acquire knowledge. This shift has opened the way to the study of scientific communities and the way in which they generate and preserve intellectual traditions. As a result, the social and the philosophical approaches to science no longer seem so widely separated, and sociologists have turned once more to the sociology of scientific knowledge. Furthermore, since both sociologists and philosophers draw many of their case studies from history and pay more attention to science in its historical context, historians are finding common ground for discussion with sociologists and philosophers.

Some steps have already been taken in that direction. During the 1960s both Kuhn and Toulmin made contributions to conferences dealing with technology, although neither was prepared to venture very far from their base in science.[53] More recently, scholars in Germany, such as Weingart and some members of the Starnberg group, have reopened the question of the relationship of science and technology, but from a Kuhnian perspective.[54] In an ironic move, Barnes, the sociologist of science, has taken the changing view of the relations between science and technology as exemplary for the new attitude he wants to foster towards the description of relationships between science and other forms of human activity.[55] Anthony Wallace, drawing on his understanding of Kuhn to illuminate cultural change, has now come to study the culture of technology under this rubric.[56] David Wojick has suggested that Kuhnian paradigms can be applied not so much to the generation of new technological knowledge, but to the assessment of it by decision-makers.[57] But perhaps the most ambitious use of theories of scientific change to illuminate technological change thus far occurs in Edward Constant's *Origins of the Turbojet Revolution* where he suggests that technological traditions are identified with easily recognizable communities of practitioners with a certain knowledge basis.[58] He hypothesizes that revolutions occur in technology when a new technological paradigm, and a different community of practitioners comes to dominate.

Revolutions are precipitated, just as Kuhn suggests they are in science, by the accumulation of anomalies – that is to say, by the accumulation of circumstances in which the technology breaks down. Constant also suggests that there is a distinct kind of anomaly, peculiar to technology, which he terms a presumptive anomaly. He defines this as an occasion where there is no direct evidence of the failure of technology, but when scientific theory suggests that in certain circumstances the technology will fail. The turbojet revolution, in Constant's opinion, was brought about by just such a presumptive anomaly.

With these developments in mind, a workshop on 'Models of Scientific and Technological Change' was sponsored by the Center for Philosophy of Science at the University of Pittsburgh in April 1981 with the purpose of examining the relevance of recent historical, philosophical and sociological studies of changing scientific knowledge for technology. A small interdisciplinary group of historians, philosophers and sociologists was invited, and the papers in this volume are revised versions of presentations made at that workshop. Sadly Derek Price's death meant that he was unable to see his paper published.

Since the workshop, economists and sociologists have been independently developing similar kinds of approach, Giovanni Dosi, for example has tried to apply Kuhn's model to technology and to use it to criticize standard economic models. Trevor Pinch and Wiebe Bijker are attempting to develop a unified theory that covers both science and technology.[59] Such work should lead to new and constructive dialogue about technology.

Issues already raised in this introduction surfaced again and again during the workshop. The relationship between science and technology was treated at length by Gutting and Price, and in passing by every participant. No one disagreed with the proposition that technology constituted a cognitive activity in its own right, and that to treat it as applied science was a mistake. Equally everyone was quite prepared to admit that, while technology contains a tacit element, this is not so large as to hinder investigation of the cognitive structure of technology. The association of particular cognitive traditions in technology with specific communities was also accepted as unproblematic, but as very much in need of further research. Constant, Weingart and Gutting all pointed out that community structures were much less well-defined in technology than in science, and Hummon offered some detailed examples which served as a basis for discussion. One source of confusion was whether the community should be defined as that group responsible for generating new technology, or whether one should look instead to the wider community bringing that technology to society. Some discussion, particularly by Weingart and Gutting, focused on the ways in which values of various communities outside technology influenced problem-selection and assessment of new technologies, a situation that they contrasted with the relatively autonomous position of the scientific community.

The *relevance* of recent work in philosophy of science to technology was discussed. Most of the participants believed that, always providing the disanalogies were kept in mind, the approach held out some promise. Price challenged this by forcefully advancing his position that the source of change in both science and technology comes not from internal developments within those areas, but from the often unexpected advent of new instrumentalities. Weingart concluded that orienting factors in technology were so different from those in science that attempts to draw strict analogies were apt to be strained. Gutting tried to keep participants honest by reminding us that if Kuhn's position was construed as amounting to nothing more than the assertion that supertheories existed then its application to almost any area of human activity became trivial.

Specific mechamisms for technological change were also discussed, by Price, as already mentioned, and by Hummon, Constant, Weingart and Laudan. A variety of driving forces were advanced – new instrumentalies, normal operating procedure within organizations, economic and social forces, and failures within the technological system itself. All agreed that this was an area that required more research.

The participants in the workshop showed sustained good will and a sense of humor in coping with the trials of interdisciplinary interaction. There was general agreement that the issues were timely ones, and a healthy willingness to lay aside disciplinary rigidity in order to engage in a shared enterprise. The discussion was always lively and sometimes heated, creating an atmosphere of concern over shared problems that is regrettably difficult to capture in an anthology. Since then I have benefitted by discussing these ideas with members of two reading groups at Virginia Tech, one on technological change, and one on history of science and technology. During these I have been stimulated by conversations with colleagues from different disciplines who have expressed incredulity about many aspects of this enterprise: that historical work could produce grounds for theory building (mainly social scientists), that theory building was a worthwhile enterprise (mainly historians) or that there was anything of interest whatsoever to discuss in technology (mainly philosophers). Such wholesale assaults on the research program made the difficulties of interdisciplinary work, where what is taken as an unproblematic cliché in one field is regarded as controversial or even outrageous in another, abundantly clear. If this anthology can go any way to building bridges between some of the disciplines that study technology, it will have accomplished its task.

Finally, I would like to acknowledge the generous support of the Westinghouse Electric Corporation, the R. K. Mellon Foundation and the Center for Philosophy of Science at the University of Pittsburgh in sponsoring this workshop, and Becky Cox who has rendered invaluable assistance in typing the manuscript.

Science Studies Center
Virginia Tech

NOTES AND REFERENCES

1. Thomas P. Hughes, 'Emerging Themes in the History of Technology,' *Technology and Culture* 23 (1979), 697-711.
2. Much of this emerging consensus has yet to be published in monograph form, and the original studies that form the basis are in articles (largely in *Technology and Culture*) and in dissertations. Perhaps the easiest way to appreciate that it exists, outside of society meetings is in the syllabi published from time to time in various newsletters.
3. For references to these works, see the notes in the remainder of the introduction.
4. The pioneering work of S. L. Gilfillan, *The Sociology of Invention* (Chicago: Follet, 1935) and *Supplement to the Sociology of Invention* (San Francisco: San

Franciso Press, 1971) has attracted few successors. For a rare exception, see the survey by Hedvah L. Shuchman, *Information Transfer in Engineering* (Glastonbury, Conn: The Futures Group, 1981).

5. For an excellent survey of economic models, particularly those applicable to history, see the editor's introduction in A. E. Musson, ed., *Science, Technology and Economic Growth in the Eighteenth Century* (London: Methuen, 1942), 1-68. Other summaries are given in Edwin Mansfield's essay 'The Economics of Industrial Innovation: Major Questions, State of the Art and Needed Research,' in Patrick Kelly and Melvin Kranzberg, eds., *Technological Innovation: A Critical Review of Current Knowledge* (San Francisco: San Francisco Press, 1978) and in several of Nathan Rosenberg's collected essays, *Perspectives on Technology* (Cambridge: Cambridge University Press, 1976).
6. For this point, see Nathan Rosenberg, *Inside the Black Box: Technology and Economics* (Cambridge: Cambridge University Press, 1983). As an example, the heated debate about railroads and American economic growth in the mid-nineteenth century is conducted entirely in external terms, and does little to illuminate the ways in which railroad engineers struggled with the technological problems facing them. For a useful summary of this debate sparked by Robert Fogel, *Railroads and American Economic Growth* (Baltimore: The Johns Hopkins University Press, 1964) see Paul A. David 'Transport Innovations and Economic Growth: Professor Fogel on and off the Rails,' in *Technical Choice, Innovation and Economic Growth* (Cambridge: Cambridge University Press, 1975).
7. See the classic papers by Moses Abramovitz, 'Resource and Output Trends in the U.S. since 1870,' *American Economic Review Papers and Proceedings* (1956), 1-23 and Robert Solow, 'Technical Change and the Aggregate Production Function,' *Review of Economics and Statistics* (1957), 312-20, and the subsequent flurry of studies on technology in the economic literature.
8. Similarly, some organization theorists look to technology as a possible determinant of the different forms of organizations. See, for example, Charles

Perrow 'A Framework for the Comparative Analysis of Organizations,' *American Sociological Review* 32 (1967), 194-208 and James Thompson, *Organizations in Action* (New York: McGraw Hill, 1967).

9. History is replete with examples where economic necessity failed to generate an adequate technological response. For example, despite the shortage of wood in seventeenth-century Britain, attempts to substitute coke for charcoal in the iron smelting process met with repeated failure, until Abraham Darby was partially successful in the early years of the eighteenth century. Examples such as these have led historians of technology to inverse the familiar slogan, claiming that 'invention is the mother of necessity'.
10. A good example of the manifold meanings of technological change can be found in the eclectic collection of readings in John G. Burke and Marshall C. Eakin, eds., *Technology and Change* (San Francisco: Boyd and Fraser, 1979). Ellul and Florman, Einstein and Bronowski, Engels and Schmookler, as well as a host of other scholars from disciplines and intellectual orientations totally remote from one another jostle for attention. Justifiable in an anthology for introductory use, this open-mindedness becomes a hindrance if more scholarly analysis is the aim.
11. See Gary Gutting, *Paradigms and Revolutions: Applications and Appraisals of Thomas Kuhn's Philosophy of Science* (Notre Dame, Indiana: Notre Dame University Press, 1980).
12. Michael Polanyi, *Personal Knowledge: Towards a Post-Critical Philosophy* (Chicago: Chicago University Press, 1958), 49.
13. See Kelly and Kranzberg, *Technological Innovation*, ch. 4.
14. This is of course not the only problem. Others include developing supporting technological and economic structures, for example. See Everett Rogers, *Diffusion of Innovations* (New York: The Free Press, 1962) for a more complete description.
15. See, for example, John Fitchen's *The Construction of Gothic Cathedrals: A Study of Medieval Vault Construction* (Chicago: University of Chicago Press, 1961), for a fascinating account of his efforts to reconstruct the methods used by medieval masons for

vault construction, or Donald Cardwell's account of the difficulties encountered in building a replica of a Newcomen engine, in *Turning Points in Western Technology* (New York: Science History Publications, 1972), 68.

16. Eugene Ferguson 'The Mind's Eye: Non Verbal Thought in Technology,' *Science* 197 (1977), 827-836. See also Derek J. deSolla Price, 'Is Technology Historically Independent of Science? A Study in Statistical Historiography,' *Technology and Culture* 6 (1965), 553-568, for an analysis of the disinclination of technologists to verbalize.
17. Jennifer Tann, ed., *The Selected Papers of Boulton and Watt*, vol. I (Cambridge, Mass: MIT Press, 1981), or Walter G. Vincenti, 'The Air-Propeller Tests of W. F. Durand and E. P. Lesley: A Case Study in Technological Methodology,' *Technology and Culture* 20 (1979), 712-751 and Stuart W. Leslie, 'Charles F. Kettering and the Copper-Cooled Engine,' *ibid.*, 752-776. These references, and those in the following footnotes are illustrative. Further examples could be found by perusing *Technology and Culture*.
18. Eric Robinson and A. E. Musson, *James Watt and the Steam Revolution* (*New York*: *Augustus M. Kelley*, 1969).
19. Francis Evans, 'Roads, Railways and Canals: Technical Choices in Nineteenth-Century Britain,' *Technology and Culture* 22 (1981), 1-34.
20. Thomas P. Hughes, 'The Electrification of America: The System Builders,' *Technology and Culture* 20 (1979), 124-161.
21. Eda Fowlkes Kranakis, 'The French Connection: Giffard's Injector and The Nature of Heat,' *Technology and Culture* 23 (1982), 1-38.
22. Nathan Rosenberg ed., *The American System of Manufactures*: *Parliamentary report by George Wallis and Joseph Whitworth*, 1855 (Reprint, Edinburgh: Edinburgh University Press, 1981).
23. Bruce Sinclair, *Philadelphia's Philosopher Mechanics*: *A History of the Franklin Institute 1824-1865* (Baltimore: Johns Hopkins University Press, 1974).
24. Merritt Roe Smith, *Harper's Ferry Armory and the New Technology*: *The Challenge of Change* (Ithaca: Cornell University Press, 1977); Anthony Wallace, *Rockdale*: *The Growth of an American Village in the*

Early Industrial Revolution (New York: W. W. Norton, 1980) and Anthony Wallace, *The Social Context of Innovation* (Princeton University Press, Princeton, 1982).

25. See Edwin Layton, 'Mirror-Image Twins: The Communities of Science and Technology in Nineteenth-Century America,' *Technology and Culture* 12 (1971), 562-580.
26. This expression was coined by the philosopher, Dudley Shapere.
27. This was pointed out by Polanyi himself. See also Harry Collins, 'The Seven Sexes: A Study in The Sociology of a Phenomenon, or The Replication of Experiments in Physics,' *Sociology* 9 (1975), 205-24.
28. Edwin Layton, 'Technology as Knowledge,' *Technology and Culture* 15 (1974), 31-41.
29. Note three recent, and typical examples of the persistence of this claim. "Technology, a synonym for experiment, is a name for the applications of science." Daniel Boorstin, *The Republic of Technology* (New York: Harper and Row, 1978), xiii; "Technology" [is] broadly defined to include all varieties of applied physical and biological science and engineering, and also basic research that might soon lead to a proposed technological development," Edward Lawless, *Technology and Social Shock* (New Brunswick, New Jersey: Rutgers University Press, 1977), 9; and "Technology is science plus purpose. While science is the study of the laws of nature, technology is the practical application of those laws toward the achievement of some purpose or purposes." Richard C. Dorf, *Technology and Society* (San Francisco: Boyd and Fraser, 1974),1.
30. See the articles by James K. Feibleman, James Kip Finch, A. Rupert Hall, Peter F. Drucker, Henry M. Leicester, Cyril Stanley Smith, Fred Kohlmeyer, Floyd Litterum, Milton Kerker and John B. Rae under the heading 'Science and Engineering' in *Technology and Culture* 2 (1961), 305-99; those by John J. Beer, Derek J. de Solla Price, Robert P. Multhauf, Carl W. Condit and Robert E. Schofield under the heading 'The Historical Relations of Science and Technology,' *Technology and Culture* 6 (1965), 547-95, and those by James K. Feibleman, Mario Bunge, Joseph Agassi, J. O. Wisdom, Henry Sholimowski and I. C. Jarvie

under 'Toward a Philosophy of Technology,' *Technology and Culture* 7 (1966), 318-90. More recent literature on the subject is summarized in Wolfgang Krohn, Edwin T. Layton and Peter Weingart, eds., *The Dynamics of Science and Technology*, Sociology of the Sciences Yearbook II. (Dordrecht, Holland: D. Reidel, 1976).

31. See numerous articles by Derek Price, including 'On the Origin of Clockwork, Perpetual Motion and the Compass,' *U.S. National Museum Bulletin* 218 (Washington, 1959), 82-112, 'Automata and the Origins of Mechanism and Mechanistic Philosophy,' *Technology and Culture* 5 (1964), 9-23, and his essay in this volume.

32. D. S. L. Cardwell, *From Watt to Clausius: The Rise of Thermodynamics in the Early Industrial Age* (Ithaca, New York: Cornell University Press, 1971), and Eda Fowlkes Kranakis, 'The French Connection.'

33. C. C. Gillispie, 'The Natural History of Industry,' *Isis* 48 (1957), 398-407.

34. *Project Hindsight* (1969), Office of the Director of Defense Research and Engineering, 2 vols. (Chicago, 1968). Illinois Institute of Technology Research Institute. *Technology in Retrospect and Critical Events in Science*, (National Science Foundation, Washington, 1969).

35. Layton, 'Technology as Knowledge.'

36. Edwin Mansfield, *Technological Change* (New York: W. W. Norton and Co., 1971), 9, defines technology as "society's pool of knowledge regarding the industrial arts." Richard Nelson, reporting on the deliberations of a group composed mainly of economists, indicates that "it was generally agreed that inventive activity was a form of problem solving," *The Rate and Direction of Inventive Activity: Economic and Social Factors* (New York: Arno Press, 1975), 7.

37. Brooke Hindle, *Emulation and Invention* (New York: New York University Press, 1981).

38. See Krohn et al., *Dynamics of Science and Technology*.

39. For surveys of these, see the compendia by Paul T. Durbin, ed., *A Guide to the Culture of Science, Technology and Medicine* (New York: The Free Press, 1980) and Ina Spiegel-Rosing and Derek J. de Solla Price, eds., *Science, Technology and Society* (London: Sage, 1977).

40. For a survey of this literature see Carl Mitcham, 'Philosophy of Technology,' in Durbin, *Guide to the Culture of Science, Technology and Medicine*, 282-363.
41. Rogers, *Diffusion of Innovations*.
42. For example, the German tradition exemplified by Werner Sombart, and American work indebted to this, such as Lewis Mumford, *Technics and Civilization* (New York: Harcourt, Brace and World, 1934) or Lynn White's classic *Medieval Technology and Social Change* (Oxford: Oxford University Press, 1962).
43. Thus Alfred Chandler's fascinating study *The Visible Hand: The Managerial Revolution in American Business* (Cambridge, Mass: Harvard University Press, 1977) and the numerous studies at the interface of business history and the history of technology that it has spawned are only tangentially relevant to our concerns.
44. This point is forcibly made by Rosenberg, 'The Machine-Tool Industry, 1840-1900,' in *Perspectives in Technology*. The case is explored at more length by David Hounshell 'From the American System to Mass Production: The Development of Manufacturing Technology in the United States, 1850-1920,' Ph.D. dissertation, University of Delaware, 1978.
45. J. A. Schumpeter, *Capitalism, Socialism and Democracy* (1942; revised ed., New York: Harper, 1950), 132.
46. John Enos, 'Invention and Innovation in the Petroleum Refining Industry,' in *Rate and Direction of Inventive Activity*, 299-321. Another classic example is Jacob Schmookler's *Invention and Economic Growth* (Cambridge, Mass: Harvard University Press, 1966).
47. Gilfillan, *Sociology of Invention*.
48. Abbott Payson Usher, *A History of Mechanical Inventions* (New York: McGraw-Hill, 1929). See also the research mentioned in footnotes 16 through 26.
49. Wallace, for example, has claimed that "the fraternity of mechanicians who invented the machinery of the Industrial Revolution [was] a small cache of men, on the order of a few hundred, who over the course of about three hundred years (roughly 1600-1900) made the great mechanical improvements that preceded the electrical age." *Social Context of Innovation*, xi.

50. Usher, for example, explicitly tried to tell the history of mechanical inventions in terms of a problem-solving model derived from Gestalt psychology, *History of Mechanical Inventions*, ch. 2. Wallace used an adaptation of the Kuhnian model to quote his research on the cotton mill town of Rockdale, *Rockdale*, appendix.
51. Thomas Kuhn, *The Structure of Scientific Revolutions*. 2nd ed., (Chicago: Chicago University Press, 1970); Imre Lakatos, 'Falsification and The Methodology of Scientific Research Programmes,' in *Criticism and the Growth of Knowledge*, ed. Imre Lakatos and Alan Musgrave, (Cambridge: Cambridge University Press, 1970); Stephen Toulmin, *Human Understanding* Vol. I, (Princeton: Princeton University Press, 1971). and Larry Laudan, *Progress and its Problems* (Berkeley: University of California Press, 1977).
52. For an account of this 'received view' see Frederick Suppes, *The Structure of Scientific Theories* (Urbana, Illinois: University of Illinois Press, 1973).
53. Kuhn commented on two papers in the session on 'Non-Market Factors,' in a 1960 conference on 'The Rate and Direction of Inventive Activity: Economic and Social Factors,' in Nelson, *Rate and Direction of Inventive Activity*. Stephen Toulmin gave a paper on 'Innovation and the Problem of Utilization' in an MIT conference on 'The Human Factor in the Transfer of Technology,' reported in William H. Gruber and Donald G. Marquis, eds., *Factors in the Transfer of Technology* (Cambridge, Mass: MIT Press, 1969), 24-38.
54. See Krohn, et. al., *Dynamics of Science and Technology*.
55. Barry Barnes, 'The Science-Technology Relationship: A Model and a Query,' *Social Studies of Science* 12 (1982), 166-72.
56. Anthony Wallace, 'Paradigmatic Processes in Culture Change,' *American Anthropologist* 74 (1972), 467-78.
57. David Wojick, 'The Structure of Technological Revolutions,' in George Bugliarello and Dean B. Doner, eds., *The History and Philosophy of Technology* (Urbana, Illinois: University of Illinois Press, 1979).

58. Edward Constant, *The Origins of the Turbojet Revolution* (Baltimore: The Johns Hopkins University Press, 1980).
59. See Giovanni Dosi, 'Technological Paradigms and Technological Trajectories: A Suggested Interpretation of the Determinants and Directions of Technical Change,' *Research Policy* 11 (1982), 147-62, and Trevor Pinch and Wiebe Bijker, 'The Social Construction of Facts and Artefacts: Strategic and Methodological Imperatives for a Unified Approach Toward the Study of Science and Technology,' unpublished m.s., 1983. See also D. Sahal, 'Alternative Conceptions of Technology,' *Research Policy* (1981).

Edward W. Constant II

COMMUNITIES AND HIERARCHIES: STRUCTURE IN THE PRACTICE OF SCIENCE AND TECHNOLOGY

The purpose of the model proposed here for technology was and is historical; that is, it is intended to be utilitarian and heuristic rather than definitive or 'clean'. It grew out of work on turbine systems, specifically turbojets, in which I found conventional explanations of invention inadequate: the turbojet is neither the product of unique genius nor a simple-minded 'new combination of old ideas'. I therefore developed an 'ideal-typical' model for the structure of technological practice, and for technological change. I use 'ideal-typical' in its Weberian sense: an artificial construct intended to portray the essential relations among conjectured entities, the purpose of which is to illuminate rather than to replicate 'social reality'.[1] The goal is not comprehensive or finished theory, but rather the beginning of secure historical understanding.

In this context, this paper is intended to do three things: first, to sketch briefly a notion of science as a basis for comparison and discussion; second, to offer a somewhat more fully developed description of technology; and third, to examine three specific areas that both represent major differences between science and technology and signal issues for each enterprise that require much greater attention. Those three areas are: (1) the hierarchical structure of technological practice and the possibly less hierarchical structure of science; (2) 'satisficing modes' in technology and science, both those explicitly sanctioned and those implicitly tolerated; and (3) the role of economic and social criteria in each enterprise which may serve as external constraints, or as internalized values or imperatives.

My notion of science is a concatenation of the ideas of a number of scholars. From Thomas Kuhn comes the cardinal notion that the cognitive locus of science is a well-defined community of practitioners, which is tautological with some 'paradigm', tradition of practice, or esoteric body

R. Laudan (ed.), The Nature of Technological Knowledge, 27–46.

of knowledge. Significantly, my ready assent to Kuhnian views probably derives from my focus on the social or historical processes of science and technology, rather than on the epistemological status of scientific knowledge.

The issue is what practitioners do, which to me is a promising and fruitful path into what they know and how it changes. One of the best short versions of what at least some scientists do and how it relates to generalizations about the pursuit of science is given by Diana Crane in her study of high-energy physics.[2] Crane addresses such central issues as problem selection, evaluation criteria, and the 'internal', presumably hierarchical, structure of scientific knowledge. In grossly simplified terms she finds that the behavior of theoretical high-energy physicists responds to a variety of hierarchically-disposed precepts, including symbolic generalizations (general principles of the field), values (such as testability and predictive power), metaphysical models, and exemplars (which Crane defines as "model[s] that [are] broad and testable with a high level of precision").[3] The image of science that emerges then is not exactly Popperian or Kuhnian or Lakatosian, but exhibits facets of each.[4]

For purposes here, two elements of this notion of science deserve special emphasis. The first concerns testability. While few scientists may really try very hard to refute their own theories, (if nothing else, getting any complicted experiment, like any complex system, to work at all is a major challenge, so the tendency is to publish and move on), the value of testability, the possibility of obtaining replicable, inter-subjectively corroborated, or, if you will, objective results remains compelling, both in evaluating results and, *ex ante*, in selecting research problems. This assertion leads to my second point. I share Donald Campbell's view that the 'locus of objectivity' in science (assuming there is any) lies in some socially-prescribed set of rituals and behavioral norms which define and enforce testability, replicability, open, candid and prompt publication, and so on.[5] Thus I view science as a socially-constructed activity, and I believe it to be objective – and I think it is both.

Technological practice has major and critical structural similarities to scientific practice as just described. First and foremost, all technological practice is dominated by well-defined communities of practitioners which are tautological with equally well-defined, well-winnowed traditions of technological practice. These communities and these traditions are the central locus of technological cognition.[6]

Furthermore, community and tradition are the primary locus of what is called technological progress: the bulk of technological change, whether defined in terms of results achieved or of effort expended, consists of incremental improvements in existing practice.[7] It is 'normal' technology in the sense that solutions to problems are sought only within the purview of the conventional system. Over long periods of time, often half a century or more, such intensive development results in truly impressive progress (examples might include steam turbines, reciprocating internal combustion engines, catalytic petroleum cracking, computer architecture).

Many, if not most, communities of technological practitioners and the broad traditions they possess are readily identifiable. Turbojet engines in this country are built by two major manufacturers (General Electric and Pratt & Whitney) and one smaller one (Garret Air-Research); gas turbine units are built by those three plus the Allison Division of General Motors and what used to be the Solar Division of International Harvester. World wide, perhaps twenty firms build gas turbine systems. In the United States, three firms build large commercial aircraft; two firms build all small aircraft piston engines. Automobiles are built by one big company, two smaller ones, and Franco-American Motors and Volkswagen. The list goes on and on. Economists call it oligopoly and are dismayed. The fact remains that virtually every high technology is dominated by a few firms who together form a readily-identifiable, highly-visible community of practitioners. This assertion applies not only to major systems, such as aircraft, but also to sophisticated component parts, capital goods, and process goods. The Standard Industrial Classification (SIC) Codes are a good guide to community structure; trade and professional associations are even better. In all cases, technical similarities within a field are much greater than differences.

Individual practitioners also partition into well-defined communities. Professional-level engineers and technicians ordinarily share both common educational backgrounds and common career experiences by which they are prepared for full community membership. In the same way as a scientist's preparation for community membership includes both formal education and apprenticeship, often as a 'post-doc' with a senior researcher, so preparation for technological community membership includes both education and 'learning by doing'. A mechanical engineer may learn the essentials of thermodynamics and even experiment with a single-cylinder laboratory engine as part of his formal education. He learns how Ford builds a Ford engine in the design shop. Thereafter he personally is species 'piston engine practitioner', variety 'Ford automobile'; in twenty years he is likely to be functionally illiterate in turbine design and uncomfortable even at Chevrolet. Technological community, then, as I am using it, applies equally well to a handful of identifiable corporate entities or to some aggregation of properly acculturated individuals. Either set (firms or persons) admits of further decomposition into smaller sub-communities. This isomorphism in the way corporate entities or individuals partition into communities and sub-communities turns out to be critical, as noted below.

Ironically, probably the best evidence for the cognitive and functional dominance of technological communities and their traditions of practice comes in episodes of abrupt transition, or technological revolution. Old communities and traditions virtually never birth radically new technologies. No manufacturer of piston aircraft engines invented or independently developed a turbo-jet. No designer of piston aircraft engines independently invented a turbojet. No designer of conventional reciprocating steam engines invented a steam turbine, and no manufacturer of reciprocating steam engines independently developed a steam turbine. No manufacturer of steam locomotives independently developed diesel engines. In the case of both firms and individuals, community practice defines a cognitive universe that inhibits recognition of radical alternatives to conventional practice. When abrupt transitions in technological practice do occur, as happens from time to time, they almost always are the work of people outside, or at least on the margins of, the conventional community.

Such radical change is usually precipitated either by 'functional-failure' or by 'presumptive anomaly'. No technology is ever perfect, and the normal course of technological development is incremental improvement – the slow, deliberate articulation or extension of the received tradition. Often this process encounters difficulty, which is usually overcome without major change. Occasionally, however, problems persist: the system cannot perform under 'new or more stringent conditions'. A classic example is the inability of unsupercharged piston engines to operate at high altitudes. Functional-failure of this sort may lead to search for radical alternatives, and to technological revolution. Functional-failure can result from system-derived demands, such as technological disequilibrium (both inter- and intra-system), and 'reverse salients in a technological front'.[8] Functional-failure also includes an especially potent mode of change deriving from properties of macro-systems: technological co-evolution. Some technologies, although belonging to independent traditions, are so closely linked in systems that they evolve together, each defining the most significant elements of the other's environment. Steam turbines and large electric generators, for example, co-evolved in electrical generation and transmission systems.[9] All functional-failure based problems or anomalies are objective: a specific system really does not work very well in a specific context.

Some technological anomalies, however, are merely presumed to exist, and these I have called presumptive anomalies. They occur when scientific theory predicts either that the conventional technology will fail under some projected conditions, or that an alternative technology would do a much better job. In these cases there is no objective or overt problem in the conventional system, which might be doing very well indeed. Science serves as a look-ahead mode, as a means for vicariously exploring future or alternative environments which the technology has not yet encountered. Turbo-jets were invented not because piston engines failed to work, but because the science of aerodynamics suggested that high-altitude, near-sonic flight would require an alternative to the piston engine and propeller. Presumptive anomaly defines a logical relationship between specific scientific theory and specific technological practice, and is relatively insensitive both to community boundary definitions and to hierarchical level.

Perception of the magnitude or severity of a crisis in technological practice, of the existence of disturbing anomaly of any sort, however, is extraordinarily sensitive to point of view, to hierarchical level. PCB transformer cooling problems, for example, may constitute a profound, and public, crisis at one level, especially for the practitioners involved, but from the higher level perspective of high-voltage transmission systems as a whole may appear only as a routine if noisome problem in incremental development. Similarly, much of the debate in science about Kuhnian 'anomaly' and 'crisis' versus Lakatosian 'critical discussion' may resolve into perceptional level differences.[10]

In addition to homologous community and epistemological structure, technological practice exhibits other critical parallels to scientific practice. Technological practice, like scientific, is governed by general, socially defined and sanctioned sets of norms and values. Perhaps most similar are the norms requiring rigorous testing, candid reporting, and replicability in experimental results. The literature of technology is extensive and abounds with detailed design and testing data.[11] The purposes of science and technology may be different and the 'product' of technology is not the scholarly paper. Nonetheless, similar normative imperatives remain: no engineer, anymore than a scientist, can get away with fudged data, obscure concepts, or imprecise, inadequately-described measurements. Moreover, broadly-held norms for technological practice generate specific traditions of technological testability. For example, a whole technological tradition of dynamometers evolved in conjunction with prime mover development.[12] These traditions of testing may include both a technology (dynamometers or some set of standard testing instruments or procedures) and normative prescriptions for use and reporting.

Yet despite these basic similarities between science and technology, there are also major differences. These differences cluster around the three interrelated sets of issues noted earlier: hierarchical structure, satisficing modes, and economic and social criteria. While treated here as analytically distinct, these issues are not so easily distinguishable in practice. The focus here, moreover, is primarily on technology, with only a comparative glance at science.

The notion of hierarchical structure in technological practice is used to capture at least three ideas. First, interface constraints within technological systems are extraordinarily rigorous and demanding – much more so, I suspect, than in science. Second, at the same time, technological design problems are hierarchically decomposable; that is, resolvable into well-defined sub-problems, in ways perhaps that scientific issues ordinarily are not. Third, these considerations imply forms of interaction among specialist communities in technology that are intense and persistant (indeed, virtually continuous) in the design process. This does not seem to be true in science. Furthermore, it may be this continual interaction among practitioners that has served to obscure the specific community basis of practice in technology.

By interface constraints within technological systems I mean simply that such systems are holistic and the pieces have to fit together. The whole is greater than the sum of the parts; a functioning engine is not just a collection of pieces. The goal is not sublime elegance over some small domain: an aesthetically pleasing combustion chamber in an engine that won't run is worthless. A design which represents perfection in no single dimension – in materials used, in satisfaction of structural theory, in minimal weight, or in any other feature may be holistically far and away the best. Design, then, requires integration of esoteric knowledge – synthesis – rather than analysis. It also requires compromise. Perhaps there is some parallel here to 'normal' science (analysis and extension) versus 'grand theoretical' science (synthesis).

Second, even though technological systems are by definition holistic, their design is usually decomposable into sometimes routine, sometimes not so routine, sub-problems.[13] In most circumstances this decomposition or parsing follows community lines or (alternatively and equivalently) boundaries of technological traditions. For example, aircraft design decomposes into major subsystems: aero-structure, engine, and avionics. Structure, in turn, breaks down into various sub-assemblies: wings, tail, fuselage, landing gear, each with its own specialized problems, constraints, and design approaches. Similarly, turbojet engine design breaks out into compressor, combustion system, and turbine design, which further decomposes to blades, bearings, lubrication system, and so on. While such decomposability clearly offers immense advantages for improvements in the total systems (since individual components or sub-systems can be modified without radically changing the total system), it does place extraordinary demands on the capacity of designers to successfully integrate all sub-systems.

Third, this constraint compels intensive inter-community interaction. Specialists in materials must understand the requirements and constraints of turbine design, and vice versa; both must cope with overall engine performance requirements in an envisioned aircraft, while the aircraft designer has to give some thought to the length of the runways the plane will use. Although each specific task (turbine blade design or engine design and manufacture, say) is the province of a highly-specialized community of practitioners who possess unique and well-winnowed esoteric knowledge, creation of a functional system requires continual integration of the knowledge of different technological communities.

This continual interaction, this demand that all design be a 'team' effort, may have served to obscure the community basis of technological practice. Also contributing to the difficulty of defining community structure in technology is the degree to which technological communities and traditions of practice often overlap a number of technologies, hierarchically both higher and lower. Some basic building blocks, such as bearings or integrated circuits, crop up in an astonishing variety of final products, although fine bearings or high-quality integrated circuits are certainly the purview of well-defined communities. Similarly, some technologies, like basic foundry techniques, are used across many industries. Conversely some firms encompass a variety of partly related technologies: GM builds autos, diesel locomotives and gas turbine aircraft engines.

While clear and distinct traditions and communities do exist in technology, then, they do not have the obvious boundaries that disciplines, or even specialties, have in science. This, of couse, may be mere appearance. Science also is possessed of an array of esoteric specialties – spectroscopy, for example – that are essential to many disciplines in the same way that foundry techniques are essential to many technologies. The intercommunal interaction so critical in technology deserves more careful examination in science. In big science, especially, it could well be that while a result from an experiment in high-energy physics might enter the great science archive under one or another theoretical rubric, the practitioners actually doing the experiment might represent a wider or even different set of specialized sub-communities, possibly clustered around specific experimental techniques.[14]

The hierarchical structure of technology is related to the different modes of satisficing in science and in technology. 'Satisficing' of course refers to Herbert

Simon's by now nearly conventional view that human problem solving processes always involve search routines that are limited in extent, narrow in conception, and sub-optimal in outcome. It is the boogaboo of 'good enough', and what I am suggesting is that 'good enough' is differently defined in science and in technology.

The reason for this difference is clear. Technology explores the environment directly, not 'vicariously' as does science.[16] Technological artifacts are subject to direct environmental elimination in ways that scientific theories are not: planes crash, engines explode, wheels fall off, toasters go berserk. For this reason, technological practitioners take epistemological liberties usually denied scientists.

First, when in doubt, engineers often can over-design. If it breaks, make it bigger. It may be that 'simplifying assumptions' or clever measurement surrogates play an analogous role in science; certainly the high repute in which those scientists capable of such legerdemain are held would suggest such talents are at a premium. But the norms governing over-design in technology and simplifying assumptions in science remain obscure.

Second, one of the major ways technological theory, or theory in engineering science, differs from orthodox scientific theory lies in what is acceptable as legitimate simplification. Even highly sophisticated engineers are content working with nineteenth-century macro-level thermodynamics; many physicists would not be. Moreover, engineers deploy a massive array of information whose foundation is empirical, rather than theoretical. Yet such information in the main is both well-formulated and well-winnowed. Certainly such engineering methodologies as parameter variation or dimensional analysis suffer no intrinsic embarrassment before more orthodox 'hypothetico-deductive' methods. These satisficing strategies in technology, together with the norms governing technological testability, are different from those in science and probably reflect the different social purposes of the two enterprises.

Two cases, one from technology, one from science, exemplify these points about hierarchical structure and satisficing modes. What I want to argue is that technology has a different 'locus of ambiguity' from science. In technology, ambiguities which affect total systems performance are not tolerated; disagreements among members of specialized sub-communities or doubts about some aspect of design must be resolved, by empirical experiment if nothing else, for the total system to work. For well-developed, that is, paradigmatic science, intra-specialty or intra-community rows take precedence; disagreements across disciplinary boundaries are ignored or suppressed.

For example, the handful of men who created turbojet engines did so on the basis of new scientific theory which suggested that much higher aircraft speeds and much higher compressor and turbine component efficiencies were possible. No theory, no practice, however, suggested either that the combustion intensities necessary for a turbojet to work could be sustained or that the requisite high-temperature alloys would be forthcoming. In fact, the relevant specialist communities in both instances denied the possibility of attaining the required performance levels. Yet gentlemanly disagreement would not do. High intensity combustion at high temperature was an absolute requisite for a functional turbojet. In defiance of the best advice, turbojet protagonists undertook their own combustion system development, and, after much painstaking empirical effort, suceeded in 'pushing back the bounds of the possible'. Similarly, by exploiting existing high-alloy steels, by cajoling alloy manufacturers to try to produce higher temperature materials, and by clever design expedients, builders of turbojets finally got usable high-temperature turbines. In each case, junctures where specialists disagreed became the locus for intensive, expensive, and occasionally acrimonious, development effort: the problems had to be solved and therefore received top priority.

Science, in my opinion, is different. For example, in 1955 a young biologist named Albert Wolfson advanced the following argument.[17] His special interest was bird migration patterns, and he discovered one species of artic tern which migrated each year from Hudson's Bay across the Atlantic to South Africa, and another which left the Bering Sea and flew across six thousand miles of empty ocean to the Hawaiian islands. Some members of the Pacific species become so bloated with fat, which is necessary fuel for the long flight, that they literally burst upon landing in the days immediately before migration.

Wolfson accepted an evolutionary theory which admitted of no massive discontinuities in behavior or in physiological adaptation, especially at the species level; only slow, gradual adaptation was possible. It was therefore not possible that his terns had suddenly struck out across the Pacific and discovered Hawaii or that they had abruptly mutated to store the huge quantities of energy necessary for the voyage.

Wolfson found the solution to his puzzle in the work of Alfred Wegner on continental drift. By combining the known phylogenetic history of bird species with a theory of continental drift over the same epochs, Wolfson could resolve his paradox and not violate accepted evolutionary principles. But the continents had to move; and the geological community was unmoved. Having fought their own century-long war over catastrophism, and having no geological evidence for continental drift, the geologists in uniform voice answered no. There the issue lay – unresolved – for another decade, until geological evidence of sea-floor spreading ignited intra-community interest in plate tectonics.

A similar story could probably be told about the current debate between physicists and astrophysicists about how many neutrinos the sun ought to produce, and why some scientists seem unable to find them. In short, in science intra-community concerns come first; that is what the structure of the disciplines and of their reward systems requires. This preoccupation does not mean, however, that there is no hierarchy of theory or of values in science. Almost all scientists accept theories of conservation of mass-energy or gravitational attraction; most share some implicit notion of legitimate explanation. But by the nature of the enterprise, hierarchical agreement and inter-community harmony do not seem as important in science as in technology.

Still, scientists, like technologists, must satisfice: they do have accepted or conventional modes of limiting search space. Current theory and the limitations and capabilities of experimental technique shape problem choice and define acceptable solutions. For example high-energy physics experiments monitored by complex computer programs designed to find the expected not only limit search space, but also represent an extraordinary example of the theory-ladeness of observation. Notions about saliency, credibility, hierarchical structure, and interdisciplinary consistency operate in science, but the precise rules for defining a satisfactory solution to a specific scientific problem are not yet clear. In technology, crudely 'working' sets the outer bound for satisficing; in science, what 'working' entails still requires fuller examination.

Finally, the role of economic and social factors in shaping what communities of technological practitioners do demands attention. Edwin Layton has tagged the basic difference in the values of science and technology as that between 'knowing' and 'doing'.[18] The compelling demands placed on technological systems which face direct environmental elimination are noted above. But the effects of differing orientations in science and in technology are more pervasive and less obvious than a difference in utilitarian commitment. I therefore want to focus on the institutional structure of the practice of technology, and its consequences for historical and maybe even for philosophical interpretation of the enterprise.

As argued above, technology's cognitive locus is in a relevant community of practitioners and in the traditions of practice which the community possesses. That is obviously not the way economists, economic historians, patent attorneys, or public policy makers see it. This perceptual paradox I think is resolvable in the following manner. In the economic theory of the firm, in historical treatments of entrepreneurship, in biography or in corporate biography the focus has been on the corporate entity. This bias is understandable, given that most technological presence in the world carries a corporate logo: Coke, Xerox, the phone company, the light company, Exxon, and of course IBM. Most evidence and records are so constructed.

What this quite reasonable focus obscures is the gross homogeneity of practice across broad industrial sectors: a car is a car is a car, and there are a lot more conceptual and technical similarities between a Honda and an Oldsmobile than there are differences. The same could be said of typewriters, computers, bath towels, wide-bodied jets, toothpaste, and bubble gum. Are our perceptions of differences merely the creatures of the evil wizards of Madison Avenue? I think not.

Two factors lend substantial insight into this apparent contradiction. First, homogeneity of practice does not imply identity, only a well-winnowed tradition which grossly matches the relevant environment. Hondas and Oldsmobiles are varieties which represent 'fine tracking' of that environment; while technically similar, the differences are important and the two varieties express, really, different appreciations of the salient factors in the environment. To these real differences, firms and consumers do respond.

This 'fine tracking' is part of what Thomas Hughes has captured with his striking concept of distinctive national 'styles' in technological practice – in the development of electrical power transmission grids, for example.[19] Furthermore, it is well established that most of what is called technological progress derives from development, from small, incremental changes, from detail improvements, in well-defined systems. Revolutions are extraordinary, and however critical, usually show up as only small discontinuities in total performance progress. This situation is consistent with the notion of well-winnowed traditions of practice being well fitted to (only slowly changing) environments, and with a good deal of what we know about satisficing and cognitive processes generally.

The second reason corporations and organizations, rather than specialized communities of practitioners, appear to be the locus of technological practice offers insight into both the social role and the social context of technology. Technology is more than a technical system: it is a social function. As social function it depends on a much wider range of environmental variables than simply the esoteric knowledge required for systems design and fabrication. Organization, for example, is essential just to integrate successfully the esoteric knowledge possessed by specialized communities. Moreover, organizational variables such as reputation, reliability, permanence, service, development potential, access, and so on are critical to the actual implementation of any technology. Thus from the point of view of engineers and of ultimate users alike, the most important social entity for identifying technological function is the relevant organization or corporation. None of this vitiates what has been said before: the locus of technology as knowledge and practice is communities of practitioners; the locus of technology as social function is organization.[20]

This argument suggests a possible strategy for integrating a number of the insights offered by other authors in this volume, some of whom have noted an ambiguity in the way I construe the community of technological practitioners to comprise either individuals or organizations (corporations or institutions). Assume that a firm is composed of a structured set of functional modules: personnel, public relations, accounting, legal, engineering and design, production, shipping, service, and so on. Neglecting for a moment embodied capital (physical plant and worker skill), tacit knowledge and the problem of informal organization, then ideally these modules are fully interchangeable. Remove the engineering and production 'black boxes' and substitute others, and General Motors becomes General Electric.

This sort of metamorphosis is precisely what happened in the turbo-jet revolution – it was the mechanism by which a piston engine company like Rolls Royce became a jet engine company – and I suspect the way in which much major technological change occurs. Whether an organization does or does not belong to a certain community of technological practitioners, and whether it does or does not adhere to a certain tradition of technological practice is determined by the intellectual DNA that is the core of its technological metabolism. Certainly the rest of the organism is important, but its identity is determined by that core – which on close examination turns out to be the set individual practitioners with specific professional commitments to specific technological traditions of practice. Thus I think the ambiguity in my definition of community is only illusory.

Finer-grained analysis of such modular organizational structures should yield Norman Hummon's distinctive organizational forms. Different technologies characteristic of different social (and perhaps technological) epochs may result in different arrangements, partition, and stacking of functional organizational modules. Similarly a dynamic sub-structure or sub-routine such as Hummon's cycle of research, development, pilot plant, and production development can replace a relatively static technological function module. Specification of the hierarchical level of technological change remains a problem, as does the determination of community interfaces and information flows, but the overall analytic distinction between technology as community knowledge and technology as social function persists.

The model sketched above also allows a great deal of variation, of fine tracking of the environment, typical of well-developed technological traditions. As noted, on a grand scale Hughes calls this phenomenon distinctive national style. The variation also exists in small scale, in fact up and down the whole hierarchy of technological practice. For example, for a quarter of a century, Ford V8 engines differed from Chevrolet's, being heavier, less fuel efficient, lower in specific output (horsepower per cubic inch displacement), and capable of lower R.P.M., as well as being well-nigh indestructable. Style – and fine tracking.

If these properties of technological practice and of organization are accepted, and combined with the notion of technology as social function then a number of seeming paradoxes vanish. I have defined two fairly restricted technological senses of anomaly: presumptive anomaly which depends solely on a relationship between technological practice and scientific theory, and anomalies of the functional-failure type, which include failure to work, technological disequilibrium, and the systemic phenomenon described as technological co-evolution. Now, assume that each of the functional modules within an organization can be thought of as having a semi-permeable environmental interface. One of the functions of the module is the tracking of specific facets of the external environment and the translation of the information received for the rest of the organization. For example corporate legal departments track changing laws and regulations and devise compliance (or avoidance) strategies for the firm; marketing departments watch developing styles and cultural fads both to facilitate sales efforts and to evolve new or modified products; strategic planning groups look at long-term demographic and economic trends in order to formulate long-range investment and product plans. Thus, anomaly for actual technological practice in the functional core of the organization can come, as Rachel Laudan suggests, from all sorts of different directions not just from the presumptive anomaly and functional failure discussed here. Technology as social function must, and does, satisfy (or satisfice with regard to) a vast array of internal organizational imperatives and external cultural contingencies.

These attempts to satisfy internal organizational and external cultural requirements are the processes which are captured by Hughes' notion of style and his focus on organizational and institutional factors (patent position, for example), by Hummon's organizational models, by Peter Weingart's 'orientation complexes' which are the external environment to which the structure of technological practice responds, and by Gary Gutting's interpretation of David Wojick's 'evaluation paradigms'. These orientations which might be called organizational, external, social, or just multivariate 'anomalies' would seem to be both different in character from the technical anomalies, both functional-failure and presumptive, discussed earlier, and at least as compelling, although unlikely to be as well specified.

These other kinds of anomalies, if anomaly is even the right word, exhibit neither the objectivity of obvious technical failure nor the logical rigor of theoretically-derived presumption. The difference devolves from seeing technology as social function rather than as a community-based knowledge system. Thus external anomalies rarely elicit much consensus, especially early on, either about their existence or urgency, or about the adequacy of possible remedies. On the other hand, technical anomalies of either sort, although evoking strong emotional responses, still usually emerge within some shared universe of discourse where at least the relevant parameters, if not their values or importance, are agreed upon. External anomalies tend to be value-laden in a different sense – often relevant parameters are not, as Wojick argues, even consensible.

The danger in these broader formulations of technological practice and change, of course, is their breadth: they lose precision by becoming too inclusive, weaving technology once again into the seamless web of history. That risk, unfortunately, is inherent in the phenomena: John Guilmartin in his study of the Mediterranean system of warfare and commerce in the sixteenth century, perhaps as well as anyone, has demonstrated the complexity of the interrelations among technology, social structure, economics, and politics.[21] What is so striking about Guilmartin's study is that he not only shows complexity (historians are good at that) but also shows the critical importance of hierarchical structure. He is able to show that macro-properties of the system (the strategic and tactical preeminence of war galleys and the stability of the political entities which employed them) are often functions of very low-level, specific technical sub-systems, such as brass cannon-founding techniques or gunpowder manufacture. Properties of holistic systems, technical or sociotechnical, are thus derivable only from the specific, structured interrelations and the specific properties of their lower-level components. It is the task of discovering or defining the complex nature of such systems, and trying to balance simplicity and richness in our models of them, that concerns and baffles us all.

This portrayal of the economic and social factors affecting technological practice, its institutional form and its systemic properties, recommends a closer look at the practice of science – at its laboratories, experimental equipment, departmental arrangements, intercommunal interactions, and reward systems – and a more determined attempt to understand what these factors say about the philosophical and epistemological qualities of science. Some analogies are already clear. Disputes about discriminating between revolutionary and normal change in science probably resolve into disputes about hierarchical level. Derek Price's concept of the critical role of changing experimental technology and instrumentation in scientific change, captured in the marvelous phrase 'artificial revelation',[22] may be the mirror image of presumptive anomaly in technological practice. This possibility raises other questions about science: what are other (external) sources of anomaly for science? Does the explosive emergence of new disciplines or subfields in science (biochemistry, physical chemistry, recombinant DNA researach) depend in fact upon some large and unwieldy array of factors, including external theoretical development, technology, laboratory environment, Zeitgeist, or other less obvious 'orientation complexes'? The issue is not how these external factors corrupt the purity of science, but rather what social and institutional forms say about what science is and what scientists know.

In conclusion, what is really remarkable is the structural similarity evinced by science and technology – which I suspect reflects a common set of cognitive processes, individual and collective. But much closer attention to hierarchical structure, to modes of satisficing, and to social roles and context can greatly enrich the understanding of both enterprises.

Department of History and Philosophy
Carnegie-Mellon University

NOTES AND REFERENCES

1. Max Weber, 'Objectivity in the Social Sciences,' in his *The Methodology of the Social Sciences* (1904; reprint, New York: Free Press, 1964).
2. Diana Crane, 'An Exploratory Study of Kuhnian Paradigms in Theoretical High Energy Physics,' *Social Studies of Science* 10 (1980), 23-54.
3. *Ibid.*, 48.
4. See Thomas S. Kuhn, *The Structure of Scientific Revolutions* (Chicago: University of Chicago Press, 1962) for the communal, paradigmatic basis of scientific practice; Karl R. Popper, *Conjectures and Refutations*: *The Growth of Scientific Knowledge* (New York: Harper and Row, 1963) for the importance of testability and attempted refutation; Imre Lakatos and Alan Musgrave, *Criticism and the Growth of Knowledge* (Cambridge: Cambridge University Press, 1970), especially Lakatos' long essay, for progressive and degenerative research programmes.
5. Donald T. Campbell, 'Objectivity and the Social Locus of Scientific Knowledge,' Presidential Address to the Division of Social and Personality Psychology of the American Psychological Association, 1969; Robert K. Merton, 'The Normative Structure of Science,' in *The Sociology of Science* (Chicago: University of Chicago Press, 1970); Ian I. Mitroff, 'Norms and Counter-norms in a Select Group of the Apollo Moon Scientists,' *American Sociological Review* 39(1974), 579-595.
6. These ideas are more fully developed in my *The Origins of the Turbojet Revolution* (Baltimore: Johns Hopkins University Press, 1980).
7. Edwin Mansfield, *Technological Change* (New York: Norton, 1980); Nathan Rosenberg, *Technology and American Economic Growth* (New York: Harper, 1972).
8. For notions of technological disequilibrium, see Nathan Rosenberg, 'Technological Change in the Machine Tool Industry, 1840-1910,' *Journal of Economic History* 23 (1963), 414-446; for 'reverse salients', see Thomas P. Hughes, 'The Science-Technology Interaction: The Case of High-Voltage Power Transmission Systems,' *Technology and Culture* 17(1976), 646-662.

9. Edward Constant, 'On the Diversity and Co-Evolution of Technological Multiples: Steam Turbines and Pelton Water Wheels,' *Social Studies of Science* 8 (1978), 183-210.
10. See note 4 above.
11. Derek J. deSolla Price, *Little Science, Big Science*, (New York: Columbia University Press, 1963) holds that the literature of technology is sparse. I disagree with that conclusion.
12. Edward Constant, 'Scientific Theory and Technological Testability: Science, Dynamometers, and Water Turbines in the 19th Century,' *Technology and Culture*, 24 (1983), 183-198.
13. Herbert Simon, *The Sciences of the Artificial* (Cambridge: Mass.: MIT Press, 1969).
14. See, for example, Walter G. Vincenti, 'The Air-Propeller Tests of W. F. Durand and E. P. Lesley: A Case Study in Technological Methodology,' *Technology and Culture* 20 (1979), 712-751; or John V. Becker, *The High-Speed Frontier* (Washington, D.C.: NASA, 1980), for the central role of experimental technology in setting aeronautical research agendas.
15. Herbert A. Simon, *Administrative Behavior* 3d. ed. (New York: Free Press, 1976).
16. Donald T. Campbell, 'Evolutionary Epistemology,' in P. A. Schlipp, ed., *The Philosophy of Karl Popper* (LaSalle, Ill.: Open Court, 1974), 413-63.
17. Albert Wolfson, 'Origin of the North American Bird Fauna: Critique and Reinterpretation from the Standpoint of Continental Drift,' *American Midland Naturalist* 53-62 (1955), 353-80.
18. Edwin T. Layton, 'Mirror Image Twins: The Communities of Science and Technology in Nineteenth Century America,' in George Daniels, ed., *Nineteenth Century American Science* (Evanston: Northwestern University Press, 1972).
19. Thomas P. Hughes, 'Regional Technological Style,' *Tekniska Museet Symposia* (Stockholm) 1 (1977), 211-34.
20. These factors may help explain such phenomena as non-price competition, price leadership, product differentiation, stratified markets, and so on.
21. John Francis Guilmartin, Jr., *Gunpowder and Galleys* (Cambridge: Cambridge University Press, 1974).

22. Derek de Solla Price, 'Philosophical Mechanism and Mechanical Philosophy,' *Annali Dell'Instituto E Museo di Storia Della Scienza Di Firenze* V (1980), 75-85.

Gary Gutting

PARADIGMS, REVOLUTIONS, AND TECHNOLOGY

1. INTRODUCTION

Thomas Hughes has recently noted that historians of technology have been shifting from their traditional concern with the interrelations of science and technology to a focus on the nature of technological development.[1] This shift, however, provides a good opportunity for philosophers of science, who have traditionally neglected the topic, to take up the question of science and technology. The reason is that an important strategy in historians' analyses of technological change has been the application of models of scientific change – specifically Thomas Kuhn's – to technology. Reflecting on such applications – e.g., on the extent of their success, on the sorts of modifications they require in the model of science – provides a good starting point for a philosophical account of the peculiar ways science and technology are both distinct and connected.

I will begin with some general comments on the essential features of Kuhn's model of scientific development and some cautions against applying it beyond the domain of physical science for which Kuhn proposed it. Then I will discuss some recent applications of Kuhn's model to the case of technology, in particular those proposed by Edward Constant and David Wojick. Finally, I will apply the results of my discussion of Kuhn's model and its application to technology to the philosophical question of the relation of science and technology. In particular, I will discuss the conflicting views of Mario Bunge and Henry Skolimowski.

2. ON APPLYING KUHN

Kuhn explicitly insisted in the *Structure of Scientific Revolutions* that the ideas he put forward were intended to apply only to the natural sciences. Indeed, he tells us at the outset that his central concept of a paradigm presented itself to him precisely as something that the natural sciences had and the social sciences lacked:

R. Laudan (ed.), The Nature of Technological Knowledge, 47–65.

> ...I was struck by the number and extent of the overt disagreements between social scientists about the nature of legitimate scientific problems and methods....Somehow the practice of astronomy, physics, chemistry, or biology normally fails to evoke the controversies over fundamentals that today often seem endemic among, say, psychologists or sociologists. Attempting to discover the source of that difference led me to recognize the role in scientific research of what I have since called "paradigms" (x).

Nonetheless, an astonishingly large number of Kuhn's readers have seen great promise in the application of his model of change in the natural sciences to other disciplines; and there is a substantial body of literature purporting to provide Kuhnian accounts of economics, psychology, sociology, political theory, history, religion, art, and education.[2] Unfortunately, most of the attempts are based on misunderstandings of Kuhn's work that render even their successes trivial.

The key misunderstanding has been to take paradigms as 'supertheories'; that is, as wide ranging sets of fundamental beliefs about the nature of some domain of reality and about the proper methods by which to study that domain. Interpreted as supertheories, paradigms are found just about everywhere. Freudian psychoanalysis, Skinnerian behaviorism, Keynesian economics, functionalist sociology, Christian theology, progressive education, among many others, all appear to be paradigms. Further, once paradigms have been discovered in a discipline, we can introduce Kuhn's notions of normal scientific practice (the development of theories in accord with the paradigm), of anomaly (explanatory failure of theories developed on the basis of the paradigm), and of revolution (replacement of a paradigm overwhelmed by anomalies with a new paradigm). Very soon it seems that, from a 'Kuhnian' viewpoint, just about any discipline has the same essential cognitive structure as do physics and chemistry; it is merely a question of settling on an accurate account of what the major paradigms are and when in the past revolutions have occurred.

Of course the very success of such applications of Kuhn is suspicious; a schema that fits so much so easily is unlikely to be very informative about anything. Moreover, the applications typically yield numerous incompatible accounts of the same discipline. In sociology, for example, a recent survey article discussed twelve different analyses of sociology in terms of paradigms, each finding from one to eight paradigms, with over twenty different paradigms noted by one or another of the analysts and no one paradigm on more than four lists.[3] Similar embarrassing riches have resulted from applications of Kuhn to economics.

The triviality of these exercises is not hard to explain. Almost any activity – surely any with cognitive pretensions – is based on some (at least implicit) general theoretical and methodological viewpoint; hence the ubiquity of paradigms construed as super-theories. Since these viewpoints underlie certain patterns of thinking and acting, they obviously ground standard modes of practice (normal science!). Further, these modes of practice – from caring for a lawn or running a business to the most sophisticated scientific theorizing – will inevitably encounter difficulties (anomalies!) that may lead us to question the adequacy of the underlying viewpoint (a crisis!) and perhaps even replace it with another (a revolution!). When paradigms are taken as supertheories, Kuhn's developmental model is a nearly empty schema that uninformatively applies to just about anything. This sort of application of Kuhn's work recalls the sterility of the numerous 'applications' of the dialectical triad of thesis, antithesis, and synthesis by Hegelian camp followers.

Kuhn's own work is not trivial because *his* paradigms are not supertheories. Rather, as he emphasizes from the very beginning of *The Structure of Scientific Revolutions*, they are model achievements (exemplars, in his later terminology); they are, that is, "universally recognized scientific achievements that for a time provide model problems and solutions to a community of practitioners" (x). A paradigm, then, is a successful piece of science that serves a community as a model for future work. Analysis of a paradigm (e.g., Newton's achievement in his *Principia*) will reveal a variety of empirical and theoretical laws, physical models, methodological rules, and metaphysical principles implicit in it. These together constitute a supertheory associated with the paradigm. But the community that accepts a paradigm may disagree substantially about the associated supertheory, and it may even be that no one has a complete and coherent formulation of it. This does not matter because the community's basic allegiance is to the paradigm as an exemplar for future scientific activities, not to any particular formulation of the supertheory.

The most important consequences of the meaning and role Kuhn gives to paradigms concern the *cognitive authority* of science. First, Kuhn rejects the idea that this authority resides ultimately in any set of methodological rules (or metaphysical principles) that govern and vindicate scientific practice. Neither the initial acceptance of a paradigm nor the normal scientific practice following on acceptance is rule-governed; so there can be no question of justifying scientific claims (ultimately) by showing that they are licensed by methodological rules. Rather, the cognitive authority of a scientific claim rests finally on the *judgment* of the scientific community that it is worthy of acceptance. A scientific claim will be supportable by appeals to methodological principles; but conflicting claims will typically be likewise supportable; and in the end scientists must simply judge that, all things considered, one claim deserves acceptance over others. The judgment in question is not that of the individual scientist. Rather, it represents a *consensus* of the scientific community. The authority of this consensus is ultimate in the sense that it derives entirely from the judgmental reliability of those who make it. Scientists, after all, are precisely those people trained to make fair and informed judgment about scientific questions. As Kuhn says, "What better criterion than the decision of the scientific community could there be?" (SSR, 169).

A bit more needs to be said about the nature of scientific consensus and of the communities in which it occurs. The consensus is about the exemplary character of an exemplar; that is, just about everybody in the relevant scientific community agrees that all their scientific work for the indefinite future should proceed along the lines suggested by the paradigmatic achievement. The most important consequence of this agreement is that the community need no longer focus on foundational questions about the nature of the domain it is investigating or about the methods of investigation that should be employed. They have a program based on implicit answers to such questions and can proceed to implement it.

Given this understanding of the way a paradigm is an object of consensus, we can readily see why there are no paradigms in the social sciences. These disciplines have produced paradigms in the sense of supertheories: comprehensive theoretical and methodological viewpoints that can serve as bases of research in a discipline. Some of these supertheories have even found a substantial following in the social sciences. But in few if any cases have they generated a consensus in Kuhn's sense. Specifically, they have not been accepted by almost all reputable members of the discipline in such a way as to eliminate the need for foundational disputes within that discipline. Kuhn was right to see paradigms (in his sense) as features that distinguish the natural sciences from the social sciences.

It is, moreover, important to note one special feature of the community in which Kuhnian consensus occurs. Since the judgment of the community itself is the ultimate basis of the cognitive authority of the claims it makes, the community has complete *autonomy* from a cognitive point of view. That is, there are no constraints on the scientific community's judgment that are not self-imposed, and the judgment cannot be appealed to any authority outside the scientific community. Lammers has plausibly suggested that this autonomy is another distinctive feature of natural as opposed to social sciences.[4]

The inapplicability of Kuhn's model to the social sciences does not eliminate the possibility that there are some disciplines and communities outside the natural sciences to which the model can be fruitfully extended. There is some plausibility to suggestions that artistic, religious, and political communities may be profitably interpreted in Kuhnian terms. Even more plausible is the thought that Kuhn's model might apply to technological communities, which seem to function in ways very like scientific communities and to achieve a similar sort of consensus about their conclusions. Let us, therefore, turn to two important applications of Kuhn's model to technology.

3. ON APPLYING KUHN TO TECHNOLOGY

Kuhn is a major influence on Edward Constant's model of technological change introduced in the first chapter of his book on the turbojet revolution.[5] Constant finds the locus of technological activity to be 'traditions of practice' that are carried on by communities of technologists. He sees these communities as engaging in the technological equivalent of Kuhn's normal science. However, although the parallel with Kuhn might suggest construing technological groups as tight-knit units of, say, engineers, Constant construes them much more broadly as hierarchical socio-economic systems that embrace a wide variety of groups concerned in different ways with technology. For example,

> the aeronautical community writ large is composed, at the least, of manufacturers, of civil and military users, of governmental and community agencies (airport authorities, for instance), and of industry, government, private nonprofit, and university-related aeronautical sciences organizations. Manufacturers, in turn, comprise communities of practitioners specializing, usually, in airframes, power plants, or accessory systems...Other highly specialized suppliers and subcontractors provide esoteric components and materials to the manufacturing organizations. Thus the aeronautical community is composed of a multilevel hierarchy of subcommunities (9).

The normal practice of a technological community can encounter difficulties (anomalies) from two very different sources. The first is 'functional failure', the failure of a technological system (e.g., an airplane) developed by normal practice to operate properly under new conditions (e.g., higher altitudes) or to meet needs imposed by improvements in other technological systems (e.g., the failure of conventional airframes to function at speeds attained by jet engines). Anomalies of this sort arise from the activities of the technological community itself or from those of related technological communities. But another sort of anomaly has an external origin in the work of pure science. This is "presumptive anomaly" which "occurs in technology, not when the conventional system fails in any absolute or objective sense, but when assumptions derived from science indicate either that under some future conditions the conventional system will fail...or that a radically different system will do a much better job. No functional failure exists; an anomaly is presumed to exist; hence presumptive anomaly" (15). An example of presumptive anomaly is the results of theoretical aerodynamics in the 1920s that showed the possibility of aircraft that could travel at speeds over 400 m.p.h. and the need for and possibility of new sorts of engine systems (turbojets) that could attain such speeds.

Continuing the Kuhnian analogy, Constant next introduces the notion of a *technological revolution*. This he defines as follows: "Technological revolution is the professional commitment of either a newly emerging or redefined community to a new technological tradition" (19). This definition is most interesting for the way it differs from Kuhn's notion of a scientific revolution. In Kuhnian science, a revolution is the rejection of one paradigm in favor of another and, in consequence, the abandonment of one system of normal scientific practice for another. Thus, Kuhnian scientific revolutions are simultaneously innovative and eliminative; they represent a fork in the road of science at which the community abandons one route of development for another. Constant's technological revolutions, by contrast, need not (and typically do not) represent an either/or choice for the technological community. They occur "when a new tradition of practice comprising a new normal technology is embraced" by *any* community of practitioners (19). This community may be a small subcommunity of the one following the old paradigm or it may be a new community drawn from several established technological traditions. In any case, the occurrence of a technological revolution is consistent with the continuance of normal technological work on the basis of the old paradigm. For example, the turbojet revolution did not mean the abandonment of piston-aircraft technology. An old technological tradition (e.g., the production of horse-drawn buggies) may become virtually extinct in the wake of a new tradition; but this is not an essential feature of technological revolutions. By contrast, the abandonment of the old paradigm is an essential feature of a scientific revolution.

In developing his model of technological change, Constant has primarily in mind situations in which we have fixed standards by which to assess the success or failure of a technological system. An anomaly occurs when a system does not function according to our standards or when we have scientific reason to think it will not so function in the future. But in either case the assumption is that we have fixed standards for evaluating how well a technological system is functioning. David Wojick's work[6] complements Constant's by focusing on cases in which technological change occurs because our standards for evaluating technological systems have themselves altered. His central concept is that of an *evaluation policy*. Such a policy comprises all of the considerations that enter into a technological community's assessment of the suitability and correctness of solutions to technological problems. Wojick distinguishes 'fact' and 'value' components of an evaluation policy; but both the former (e.g., scientific theories, engineering principles) and the latter (e.g., laws, court decisions, moral precepts) function as standards that must be met by proposed solutions. Every evaluation policy requires a set of *evaluation procedures* that provide methods of implementing the policy in particular cases. Given an evaluation policy and appropriate implementation procedures, decisions about the appropriateness of particular technological proposals (building a new dam, employing a new food additive) are relatively unproblematic: "...only the details unique to each case need to be worked out....The correct performance of a particular task is...a matter of good judgment within the standard procedure" (240).

Constant's and Wojick's approaches are clearly complementary. Constant presupposes an evaluation policy and procedures and focuses on the process of developing new technological systems that will conform to them. Wojick presupposes new technological systems and focuses on the process of developing evaluation policies and procedures to assess them. Each viewpoint corresponds to a major aspect of technological activity. We will go wrong if we fail to recognize that technology involves both the development and the evaluation of technological systems.

Just as Constant applies Kuhn's model of scientific change to the development of technological systems, so Wojick applies the model to changes in evaluation policy. He sees evaluation policies as paradigms that establish normal systems of technological evaluation. Anomalies arise in two ways. First, it may happen that "repeated attempts using standard procedures, fail to eliminate known ills" (244). Of course, the initial reaction to such failures will be to blame our own lack of technological ingenuity and skill. But eventually it may occur to us that there is something flawed in the standards by which we are selecting technological systems to solve a problem. Second, "new scientific or technical knowledge [may] enable us to see that our standard procedures do not evaluate all factors correctly" (244-5). Unlike the first sort of anomalies, these often arise from fields "outside the...discipline in charge of the technology in question" (245). For example, developments in ecology posed problems for the evaluation policies governing the construction of dams.

In one way disputes about a paradigmatic evaluation policy are similar to disputes about scientific paradigms: the evidence in favor of a change in policy is often 'soft' (e.g., tentative, controversial, merely qualitative) and proponents of the old paradigm emphasize the need to further articulate the new paradigm (i.e., develop the working procedures that will apply it) before it can be properly evaluated. However, in contrast to the scientific case, the decision in a dispute about evaluation policy may be forced by an appeal beyond the technological community in which it arises. Specifically, rather than making their case by continuing successful articulation of their paradigm, proponents of an evaluation policy may simply seek the endorsement of public bodies (popular opinion, legislatures) for their basic principles of evaluation. In such cases, a revolution in evaluation policy may be imposed on a technological community from the outside (e.g., by law). In the wake of such revolution, the technological community will be required to develop a set of evaluation procedures to fit the new evaluation policy.

Let us now turn to the question of the extent to which Constant and Wojick have in fact taken over the essentials of Kuhn's model in their accounts of technological change. As we saw in part 2, three features are most central to Kuhn's account: (1) the interpretation of paradigms as exemplars; (2) community consensus as the ultimate basis of cognitive authority; (3) the cognitive autonomy of the community.

(1) Neither Constant nor Wojick place any special emphasis on the idea that normal technological practice is based on the acceptance of an exemplar. Neither says anything excluding such a view, but they do not exploit the possibility. Wojick, for example, defines a paradigm as "a system of concepts...built up from and depending upon: (i) theoretical laws; (ii) classic experiments; (iii) methodological principles; (iv) an ontology of theoretical principles; (v) metaphysical principles" (242). This presents a paradigm more as a supertheory than as an exemplar, though we can imagine technologists coming to think in terms of the relevant system of concepts primarily by following an exemplar as a model rather than explicitly believing a large set of theoretical, empirical, methodological, and metaphysical principles. Similarly, Constant describes technological traditions of practice as composed of "relevant scientific theory, engineering design

formulae, accepted procedures and methods, specialized instrumentation, and, often, elements of ideological rationale" (10). Once again, the emphasis is on the paradigm as supertheory, though Kuhn's view of it as an exemplar is not excluded.

It is unfortunate that Constant and Wojick neglect Kuhn's concept of an exemplar, since it seems likely that the notion could contribute a lot to our understanding of technology. One instance is technological practices that exist independent of theoretical science (arts and crafts). In contrast to the common view that such practices are entirely unscientific, being at best instances of knowing how rather than knowing that, an analysis in terms of exemplars suggests that both the skilled artisan or craftsman and the pure scientist are in essence people who know how to adopt and extend previous exemplary achievements to new cases. Similarly, such an analysis might explain the fact that, even in cases where technology is thoroughly informed by theoretical science, particular *inventions* (concrete technological achievements) are not just the end-product of technology as applied science but themselves play a major role in stimulating further technological advances.[7]

(2) On the other hand, in the case of technology – by contrast to that of the social sciences – the construal of paradigms as supertheories rather than exemplars does not vitiate the enterprise of extending Kuhn's model. For technological communities clearly do exhibit the sort of consensus that Kuhn's model requires, so that analysis in terms of supertheories does not cover over, as it does for the social sciences, the lack of an essential condition for normal scientific practice. Nonetheless, technological consensus is not the same as scientific consensus. This shows up in both Constant's and Wojick's accounts. For Constant, there is community consensus about a new paradigm, but this consensus need not replace the consensus generated by the old paradigm. It may be located in a new technological community that merely takes its place alongside the old one. For Wojick consensus about evaluation policies may fail to arise from within the technological community but may be – indeed, may have to be – imposed from outside. In both cases, the differences in scientific and technological consensus derive from the different ways the values of science and of technology are related to the values of the wider community. We will return to this topic in our final section.

(3) Community autonomy is the one central feature of the Kuhnian model that does not transfer to technology. This is not only true, as we have just noted, of a technological community's evaluation policies; it is also true of judgments about technological systems made on the basis of a fixed evaluation policy. For, as Constant has shown in his valuable discussion of presumptive anomaly, such judgments can be decisively affected by information from 'pure' scientific communities not included in the technological community. This lack of autonomy in technological communities is also reflected in Constant's view of the technological community as a hierarchical array of many diverse communities. Such a view can verbally maintain autonomy, since all influences on technological development can be included by definition in this extremely broad community. But the need for recourse to such a patchwork construal of the technological community itself shows decisively that none of the concrete communities that make specific judgments on technological issues enjoy the autonomy of parallel scientific communities.

4. THE RELATION OF SCIENCE AND TECHNOLOGY

In this final section I propose to apply our discussion of Constant's and Wojick's extensions of Kuhn's model to the question of the relation of science and technology. Specifically, I will examine the central question, "Is technology applied science?," taking as a focal point the conflicting views of Mario Bunge and Henryk Skolimowski.[8]

Bunge offers a clear and vigorous defense of the view that technology is applied science. He formulates the distinction of pure and applied science as follows: "The method and theories of science can be applied either to increasing our knowledge of the external and the internal reality or to enhancing our welfare and power. If the goal is purely cognitive, pure science is obtained; if primarily practical, applied science" (19). Bunge recognizes that there are efforts at "enhancing our welfare and power" that are not applications of scientific theories or methods; these are arts and crafts, which, he holds, embody knowing-how (instrumental knowledge) rather than knowing-that (scientific knowledge). (As I noted above, proper appreciation of paradigms as exemplars in science may undermine this way of distinguishing the arts and crafts from science; but this is not a crucial point for our present purposes.) Bunge argues that knowing-how neither entails nor is entailed by knowing-that and so insists on a sharp distinction between arts and crafts (e.g., Roman engineering) that exist independently of theoretical knowledge and the sciences and technologies that attain and apply such knowledge.

There are, according to Bunge, two importantly different ways that technology is applied science. In some cases, technology applies scientific *theories*. The result is what Bunge calls a "substantive technological theory"; e.g., the application of the pure scientific theory of fluid dynamics yields a substantive technological theory of flight. In other cases, we encounter problems that cannot be solved by the application of any available scientific theory. In such a case, we must employ the *methods* of science to develop *de novo* a technological theory. The result is what Bunge calls an "operative technological theory". An example is a "theory of airways management" (i.e., "a theory concerning the optimal decisions regarding the distribution of aircraft over a territory" (21)). Operative theories are scientific theories, developed by technologists, about human actions.

For Bunge both substantive and operative technological theories reflect the essentially practical concern of technology in being "concerned with finding out *what ought to be done* in order to bring about, prevent, or just change the pace of events or their course in a preassigned way" (23). By contrast, pure scientific theories are "limited to accounting for what may or does, did or will happen, regardless of what the decision-maker does...." (23). The fact that technology, in contrast to science, has a practical orientation implies that science grounds technology and not vice versa. Pure theory, Bunge maintains, can ground successful practice; that is, we are warranted in moving from the truth of pure scientific claims about nature to technological claims about how to control nature for our purposes. But the converse inference – from successful control of nature by the application of a theory to the truth of the theory as a description of reality – is not legitimate: "the practical success or failure of a scientific theory is no objective index of its truth value" (24). Thus, the epistemological relation of pure science and technology is asymmetrical; science grounds technology but not vice versa.

Bunge's analysis is plausible as an account of *normal* technological practice. Within the limits of such practice, any technological achievement can be understood as either an application of some scientific theory or, at least, the employment of scientific method to solve a practical problem. Normal technology, we may agree, is applied science. But what about technological revolutions? Do they result simply from the application of scientific theories or methods? The answer is, perhaps, yes for the sort of revolutions discussed by Constant: developments of new technological systems that meet the standards of a given evaluation system. But for the cases of revolution in evaluation policy itself, discussed by Wojick, the view of technology as applied science is not generally adequate. It will be adequate to the limited extent that evaluation policy is a function of theoretical scientific commitments. But other aspects of evaluation policy (e.g., all the 'value' aspects) do not result from the application of scientific theories and methods. To take one of Wojick's examples, the revolution in the evaluation policy used for judging dam proposals that occurred as a result of the National Environmental Policy Act was obviously not a result of any application of scientific theory or method.

It may be objected that, properly speaking, technology is concerned only with the development of technological systems, not with the values relevant to decisions about employing these systems in particular situations. Technology, it might be urged, is concerned with the means by which given ends may be best attained; it has nothing to say about which ends are appropriate. Bunge himself seems to favor this sort of response, since he says that technologists "are bound to devise and implement the optimal means for achieving desiderata which are not usually chosen by themselves; they are decision makers, not policy makers" (34). Given this view, properly *technological* revolutions would be just applications of science. The value changes that do not fit Bunge's account would not be technological changes.

Obviously, we can and should conceptually separate questions of ends from questions of means. The issue, however, is whether this separation is mirrored in the communities concerned with technology. Is it plausible to distinguish one sort of community – technologists properly speaking – concerned only with questions of means, from another sort – government agencies, the general public, etc. – concerned only with ends? I think the answer is no because in practice there is an inextricable connection between what we think about ends and what we think about means. On the one hand, the concrete ends we desire depend on our beliefs about what is achievable. No sane person desires the ability to fly to the moon by his own powers or to possess all the gold on and in the earth. We must distinguish wishful thoughts from operative desires and realize that the latter vary significantly with what we know (or think we know) about available means. On the other hand, our knowledge of means depends greatly on our ends. We do not take the often considerable trouble to determine how an end can be attained unless we are interested in attaining it. For a long time, for example, there was little desire to conserve oil and so little knowledge of how to do it. As a result, technologists pursue the lines of research they do largely because they share the values of the wider community that benefits from their work; and the desires of the wider community depend importantly on the instrumental possibilities developed by technologists. Accordingly, the idea that the distinction of means and ends is mirrored in the distinction of technologists from the wider community is not viable.

The essential point can also be put in this way: although there is a sharp distinction between the questions, "Is M the best means to E?" and "Is E a good end?", there is no correspondingly sharp distinction between the questions, "Should I develop M as a means to E?" and "Should I desire E as an end?" In the latter case, a positive answer to either question requires a positive answer to the other. Since technology is the *practical* enterprise of developing means to ends and not just the speculative exercise of theoretically determining which means would best attain which ends, it is concerned with the latter pair of questions and therefore cannot be treated as merely concerned with means. It is precisely because technology is a *practical* and not a theoretical discipline that it must take account of both ends and means.

This conclusion is supported by (and also helps to explain) the lack of autonomy in technological communities that emerged in both Constant's and Wojick's attempts to apply Kuhn's model. For any community that functions as a coherent whole, there are some values that ultimately govern its actions. In the case of science, the relevant values are such things as predictive accuracy, theoretical fruitfulness and simplicity (or, more exactly, theories that exhibit these features). These values, of course, are not values of the scientific community alone. Science is supported by the wider community because the latter endorses the former's main defining values. Society as a whole wants good scientific theories. However, although scientific values are not peculiar to science, scientific communities are the exclusive custodians and arbiters of these values. That is, they alone decide how the values are to be interpreted in particular cases and whether they are fulfilled by a given enterprise. The autonomy of science consists in this exclusive authority (delegated by the wider community) over the interpretation and application of its values. The wider community allows this autonomy because pure scientific values are only indirectly connected to its most pressing concerns and because pure science seems to make a more effective contribution to these concerns when left to its own devices. By contrast, the values of technological communities are among those most immediately and critically related to one of our society's main concerns: the continual maintenance and increase of the material well-being of its members. As a result, technologists find themselves by no means alone in their immediate interest in the question, "Should we develop M as a means to E?" This question directly connects with the most immediate concerns of the wider community, and the technologist's attempts to answer it must interact with other parts of society.

Given then that the technological community should not be construed as concerned merely with means to ends but with the ends themselves, we must include Wojick's revolutions in evaluation policies as properly technological revolutions. But then, since such revolutions are not merely matters of applying scientific theories or methods, it follows that technology is not just applied science.

Skolimowski also rejects the view that technology is applied science ("it is erroneous to consider technology as being applied science" (73)); moreover, he thinks that the difference between science and technology are most apparent when we look at their development. Specifically, "the difference between science and technology can be best grasped by examining the idea of scientific progress and the idea of technological progress" (75). Scientific progress, Skolimowski claims, is a matter of gaining *knowledge* – truth about the world – and so science is primarily concerned with investigating reality. By contrast, technological progress is a matter not of gaining more knowledge but of producing better artifacts. The concern of technology is not to "*investigate* the reality that is given" but to "*create* a reality according to our designs" (Skolimowski's emphasis). Further, the goal of technological creation – the sense in which technology aims at "better" artifacts – is greater *effectiveness*; i.e., more efficient action. Skolimowski thinks that a general theory of efficient action is available in the praxiology of Tadeusz Kortarbinski; but he also sees a need for specific accounts of the special sorts of effectiveness aimed at by specific areas of technology. Thus, he suggests that surveying aims at accurate measurement, civil engineering at durability of construction, and mechanical engineering at the efficiency (in the technical sense) of engines. Further, each sort of effectiveness corresponds to specific patterns of thinking that are characteristic of each area of technology.

Although our reflections in this paper support Skolimowski's claim that technology is not applied science, they do not support his account of why this is so or of what the nature of technology is. Skolimowski holds that science but not technology has knowledge as a goal. Technology's goal is efficient action, and scientific knowledge and methods are at best employed as means to this goal. A first difficulty here is that it is not at all clear why this point, even if correct, shows that technology is not applied science. An 'application' is precisely a case of employing certain methods or results to achieve an end that may be quite different from that for which they were initially developed. Thus, much of contemporary physics is applied group theory, even though the purpose of contemporary physics is quite different from that of the pure mathematics that developed group theory. So even if we agree that science has an essentially cognitive end and technology does not, why should we conclude that technology is not applied science?

Another difficulty arises from Skolimowski's implicit assumption that because technology acts for practical ends it has no distinctive cognitive content; i.e., there is no distinctively technological knowledge. But even granting that, in contrast to science, the ultimate goal of technology is not knowledge, it does not follow that, in the course of attaining its noncognitive goal, technology might not generate a distinctive body of knowledge. In fact,

Constant's account of technological change points to one source of such knowledge: the functional failure of technological systems. The mere fact that a system fails to perform properly in certain circumstances in itself constitutes a piece of knowledge essential to the technological enterprise and is often knowledge not available from scientific theories. Moreover, Constant shows how technologies set up systems of technological testing that are specifically designed to generate knowledge of this sort. Such testing represents an application of scientific methods – Constant has especially in mind the Popperian method of conjectures and refutations – to obtain technological knowledge. As Constant says:

> It may well be that the application of this method of bold total-systems conjecture and rigorous testing...to large-scale, complex, multilevel systems beginning in the nineteenth century created a fundamentally novel category of knowledge and knowledge processes distinct both from science proper and from craft technology (21).

Finally, our reflections on technological change raise serious doubts about Skolimowski's view that efficient action is the fundamental goal of technology. Here Skolimowski seems to join Bunge in seeing technology as a purely instrumental discipline, producing effective means with no concern for the ends they are designed to implement. Thus, surveyers aim at accurate measurements and civil engineers at durable structures, with no concern for questions about the values the artifacts they produce may enhance. But, as we have seen, this sort of attempt to isolate technological communities from the values of the wider community is inadequate to the practical nature of the technological enterprise.

The results of our discussion of Bunge and Skolimowski are not only negative. They also suggest a positive view of technology and its relation to science. This view has two main elements: (1) Technology is (like pure science) a cognitive enterprise, producing its own distinctive body of knowledge about the world. (2) Technology is also (unlike pure science) a *practical* enterprise, concerned with the most immediately pressing needs of the society in which it exists. In sum, technology is neither just applied science nor just a set of techniques (of no cognitive significance) for implementing goals.

Rather, it is a body of *practical knowledge*, informed by and in the direct service of the immediate needs of the human community. In this paper, I have arrived at this view by reflection on recent work in the history of technology. I think that the same view could be reached from an epistemological direction through a suitably modified version of Habermas' idea that human knowledge is constituted by cognitive interests, including an interest in the instrumental control of nature. But this, of course, is a topic for another occasion.

Department of Philosophy
University of Notre Dame

NOTES AND REFERENCES

1. Thomas Hughes, 'Emerging Themes in the History of Technology,' *Technology and Culture* 20 (1979), 706.
2. For a bibliography of this literature, see Gary Gutting, ed., *Paradigms and Revolutions*: *Applications and Appraisals of Thomas Kuhn's Philosophy of Science* (South Bend, Indiana: University of Notre Dame Press, 1980).
3. D.L. Eckberg and L. Hill, Jr., 'The Paradigm Concept and Sociology: A Critical Review,' *American Sociological Review* 44 (1979), 925-937. Reprinted in Gutting, *Paradigms and Revolutions*.
4. C. Lammers, 'Mono- and Poly-paradigmatic Developments in Natural and Social Science,' in R.D. Whitley, ed., *Social Processes of Scientific Development* (London: Routledge and Kegan Paul, 1974), 123-147.
5. Edward Constant, *The Origins of the Turbojet Revolution* (Baltimore: Johns Hopkins University Press, 1980). Page references will be given in the text.
6. David Wojick, 'The Structure of Technological Revolutions,' in G. Bugliarello and D. B. Doner, eds., *The History and Philosophy of Technology* (Urbana: Illinois, University of Illinois Press, 1979). Page references will be given in the text.

7. Very helpful discussions with Thomas Hughes at the Pittsburgh Conference have suggested to me that the concept of an exemplar may also be very useful for understanding the process of technological discovery. For example, Hughes suggested that much of Sperry's work employed the negative feedback mechanism as an exemplar. However, these same discussions have made me realize that the notion of an exemplar requires very careful articulation if it is to be faithful to the complexities of technological thinking.
8. Mario Bunge, 'Technology as Applied Science' and Henryk Skolimowski, 'The Structure of Thinking in Technology,' both originally published in *Technology and Culture* 7 (1966), 329-347 and 371-383. Page references (given in the text) are from reprints of these articles in Friedrich Rapp, ed., *Contributions to a Philosophy of Technology* (Dordrecht, Holland: Reidel, 1974.)

Norman P. Hummon

ORGANIZATIONAL ASPECTS OF TECHNOLOGICAL CHANGE

1. INTRODUCTION

This paper discusses the organizational dimensions of two kinds of technological change. The first concerns how modern industrial organizations manage technological change, and the second concerns the development of new forms of organizations and how these new forms appear to be connected with the development of a new technology.

The words 'technology' and 'technological change' are found frequently in the organizational literature. However, the meaning attached to these words differs from the way they are used by other authors in this volume. Organizational theory has not focused on the production and use of technology, *per se*. Instead, not surprisingly, it has focused on the interrelations of social roles and jobs that are structured by technology.[1] Technology is most commonly treated as a set of conditions that constrain and influence social organization.

Another disparity exists between the way technological matters are treated in this volume, and in organization theory. Very broad classifications are used by organizational theorists. Thompson's classification is one of the most often cited and employed.[2] According to Thompson, some technologies are 'long linked' in that tasks is performed in a technically determined sequence. Other technologies are 'mediating' so that an array of tasks is performed in parallel. Finally, some are 'intensive' with mutually interdependent tasks. To give examples, an assembly line employs a long linked technology, a bank uses mediating technology, and operating surgery uses intensive technology. While these are useful distinctions to make if one is interested in trying to explain organizational structure, they are less satisfying for studying technological change. For example, the long linked category would include almost all manufacturing technology, from the small, custom job shop to the huge assembly facility in the automobile industry. While these technologies

R. Laudan (ed.), The Nature of Technological Knowledge, 67–81.

share sequential task structure, the methods used to create new job shop or assembly line technologies are different. To examine technological change in organizations, it is necessary to look more closely at specific technologies, and at the risk of losing generality, at specific types of organizations.

Much of this paper is based on personal experience and inquiry. Where it is possible to support observations with references from the research and scholarly literature, it has been done. Thus, many of the ideas offered must be considered speculative; they have not been subjected to the test of systematic research.

2. TECHNOLOGICAL CHANGE AS STANDARD OPERATING PROCEDURE

When technology is considered as an organizational *activity* rather than as a set of organizational constraints, a different picture emerges. Technological change appears to be an inherently organizational social process. While single individuals invent technologies, it is seldom the case that an individual can personally implement a new technology by, for example, bringing a new product to market. Let us examine the nature of technology and the basis of this organizational linkage more closely.

First, technology is one of the products of human rationality. It results from the rational problem solving effort required to produce a means with which to meet some societal goal or need. Second, technology extends the limited biological or physical capabilities of human beings. Tools enable people to cope more effectively with both the natural and man-made environment. Whether the tool is a hammer, a bulldozer, or a digital computer, the use of that tool improves the ability to do something, i.e., to meet some goal. While people can break rocks and move earth without hammers and bulldozers and can process information without computers, the appropriate tool augments our limited physical and mental capacities to achieve goals and meet needs with more satisfying results. Finally, technology is both a cause and a consequence of social organization. The production and the use of technology almost always involve the division of labor, whether we are talking about the two-man buck-saw or the automobile assembly line. Each of these sample technologies involves a sequence of technologically defined tasks which is mirrored by a set of social roles, and the interaction and performance of these tasks and roles defines the organizational attribute of technology.[3]

Further elaboration of these technological attributes reveals more about the importance of organization. Consider, again, rational problem solving.

3. RATIONAL PROBLEM SOLVING

Technology, as a mode of rational problem solving, draws upon two kinds of knowledge. The first is the formal, structured, body of scientific knowledge. Prime examples are the use of Maxwell's equations in electrical engineering and simplex algorithms in industrial engineering. Since World War II, this form of rational problem solving has been the almost exclusive emphasis of engineering education in this country.

A second mode of rational problem solving is heuristic and experiential in nature. In industry, it is called know-how. This mode of rational problem solving is labelled the art of technology. The importance of this mode for the successful development and implementation of technology cannot be overemphasized. This can be the case, even for the most advanced 'high technologies'. Let me provide an example based on the design of modern aircraft.[4] During this process, the functional requirements of the design are translated into physical specifications, systems, and sub-systems. Then, teams of engineers using some of the most advanced computer facilities in the world analyze the structural features of the air frame, and the aerodynamic properties of fuselage, wings, etc. However, as advanced as these scientifically-based methods are, they yield only partial, approximate design solutions. The task of solving a complete set of Namier-Stokes equations in three dimensions for an aircraft design is still well beyond the scientific mode of problem solving. And yet, a complete design requires the synthesis of all the pieces. Major contributors to this synthesis are the highly skilled craftsmen known as model makers. How well they heuristically apply their knowledge and experience in building test models often determines the overall success of a design. Similar descriptions of the design process in several other areas of 'high technology' engineering could be offered.

Another example that illustrates the importance of the art of technology is derived from a survey of how multinational corporations transfer technologies to enterprises in other countries.[5] It is very rare for technology transfers to be made without a technical assistance agreement. The purpose of this agreement is to transfer the know-how necessary to use the technology to the recipient company. In short, the agreement attempts to ensure transfer of the art of the transferred technology as well as the technology itself. Indeed, respondents to the survey ranked the quality and extent of the company's know-how of greater importance in the pricing of technology than patents, trademarks, and other company characteristics.

Taken together these two illustrations demonstrate the functional role of the art of technology. A gap almost always exists between the knowledge required to synthesize or use or produce a technology, and the knowledge that is available from the science of technology. This gap is the difference between the world of analysis and the world of action.

The importance of the art of technology in the technological change process suggests an organizational problem. How do engineers and others engaged in the design and production of technology acquire knowledge and skills in the art of technology? It is certainly not taught in engineering schools. The answer, of course, is that organizations attempt to train people to be good XYZ Company engineers, by developing procedures that accumulate collective experience and pass it on to those less familiar with the company's organizational ways. Also, it is probably the case that the emphasis placed on project teams is, in part, a company's means of 'socializing' engineers to its art of technology.

4. ORGANIZATIONAL ATTRIBUTE

These points are further amplified when one considers the division of labor associated with the development of a new product and technology. Again an example makes the point.[6] The product development cycle in the chemical industry consists of four distinct organizational phases: the research laboratory, the development laboratory, the pilot plant, and the full commercial production facility.

The research laboratory conducts scientific research on theories and methods of producing a new chemical compound. The operational mode is scientific research, the same set of activities that would be pursued in a university or government laboratory. The output of the research laboratory is a research report which describes the theory and methods, as well as the properties, both desirable and undesirable, of the new chemical compound. The *scale* of the chemical technology in the research laboratory is small; the reactions are generated in beakers, test tubes, etc.

In the next phase, the development laboratory has three tasks. The first is to review the scientific quality of the work of the research laboratory. The second is to scale up the chemical technology by at least an order of magnitude; methodologies are developed and tested to generate chemical reactions in containers on the order of five gallons. The third task of the development laboratory is to write a report that makes recommendations to the pilot plant on how to scale up the chemical technology by at least another order of magnitude. In the chemical industry, this is often an important problem. A viscous substance that was easy to stir and heat quickly in a small beaker in the research laboratory may be an extremely difficult challenge in the pilot plant in a fifty to five hundred gallon reactor and almost an impossibility in commercial production in a fifty thousand gallon pot.

The tasks of the pilot plant involve developing and testing methods to further scale up the technology, and to assess technical and economic *qualities* of the chemical reactions. The technical quality refers to the percent conversion of inputs into final products, the purity and stability of a compound and increasingly, the chemical nature of the by-products. The economic assessment focuses on the commercial viability of a new product and often indicates where further research and development would have greatest benefits.

The major managerial decisions concerning the new technology and product are made between the pilot plant and the commercial production phase. At this point, all technological, economic, and other business aspects of the new technology are reviewed. In most instances, the sum of all the resources the company invests in new technology to reach this critical point are a fraction of the total resources the company will have to allocate to go into commercial production. Thus, these decisions are crucial to the overall technological and economic success of a chemical company.

If a decision to proceed is made, the final phase of the product development cycle is implemented by the technological elite of a chemical company. A specialized team takes over a production facility. Using the reports and guidelines of the pilot plant phase, the team learns how to make the new chemical product at a commercial scale. Experience, trial and error, and large scale experimentation guide this phase of the operation. This process can go on for periods of six or more months. When the team learns how to produce the desired chemical product economically, the plant's regular management is brought back into the plant and training begins. Once plant managers and engineers are trained, production personnel are taught how to monitor, service, and in general, run a production facility. The specialized teams often retain managerial control for two years before turning the plant over to regular management.

This description of the product development cycle in the chemical industry shows the complexity of the organizational aspects of *normal technological change*. First, it is important to observe that this cycle often spans twelve to twenty years.[7] Formal organizations are more likely than individuals or informal groups to maintain the continuity of resources for the development of a new technology over such long periods of time. Furthermore, each phase of the product development cycle occurs in a different part of the organization. Typically, this means that at least two and up to four different corporate divisions are involved in the cycle. Interdivisional coordination and communication is never easy, particularly if communication channels are used that run up and down the organizational (bureaucratic) hierarchy. Some companies open 'horizontal' channels that bypass the hierarchy to improve the transfer of technological information.[8]

It is also interesting to map the relative importance of the science and the art of technology along product development cycle. Of course, the science of technology is dominant in the research laboratories, but at each succeeding step its relative importance is decreased. Conversely, the art of technology is relatively unimportant in the research laboratory environment and is all important in the final phase of commercial implementation. It appears that one of the most difficult organizational problems is the appropriate mixing of art and science and technology at each phase of planned technological change. This 'mixing' involves people, and the people who are adept at the science of technology are often quite different from those skilled in the art of technology. This stems from their different educational and organizational experiences, and perhaps most importantly, from their different views on what should be done. A failure to achieve balance causes one of two problems. Too little science results in technological stagnation and to stand still is often to fall behind. Too much science leads to paralysis of another kind; that of analysis and failure to implement. However, of the two problems, the latter has much greater consequence because technology has little value unless it is put to use.

Many industries have developed standard operating procedures to generate technological change. Moreover, these procedures and the organizational forms that implement them are quite similar within groups of related industries. All automobile companies use a similar product development cycle, for example.[9] Other groups include consumer durable companies, heavy manufacturing, electronics, pharmaceutical companies, and so on. There are, however, some very interesting differences between and across these groups.

The second part of this paper examines the organizational dimensions of technological change by considering differences in the organizational forms used to manage change. An historical approach is useful in examining these organizational forms.

5. THE DEVELOPMENT OF NEW ORGANIZATIONAL FORMS

In 1965 Arthur Stinchcombe published a paper in which he argued that:

> organizational forms and types have a history, and ... this history determines some aspects of the present structure of organizations of that type. The organizational inventions that can be made at a particular time in history depend on the social technology available at the time. Organizations which have purposes that can be efficiently reached with the socially possible organizational forms tend to be founded during the period in which they become possible.

> Then, both because they can function effectively with these organizational forms, and because the forms tend to become institutionalized, the basic structure of the organization tends to remain relatively stable.[10]

The concept of organizational form, as used by Stinchcombe, describes classes of organizations (often associated with functions) on the basis of their most important attributes such as size, vertical and horizontal differentiation, and the training or skill of their members. For example, organizations in the automobile industry tend to be very large, vertically integrated, hierarchically-structured bureaucracies. Organizations in solid state electronics tend to be flatter, highly differentiated horizontally, and they often employ a matrix structure. Machine shops tend to be small, quite informal organizations. The phrase, social technology, refers to phenomena such as the rise of literacy, the development of legal systems and markets. As an example of an organizational form, Stinchcombe points out that:

> the present urban construction industry, with specialized craft workers, craft-specialized subcontractors, craft unions, and a relation of contract between the construction enterprise and the consumer was developed in European cities before the industrial revolution.

Stinchcombe argues that this new form required the following social developments: dense settlement, contracts in law, separation of occupational roles from family roles, and a free wage labor market. These conditions do not, in general, exist in pre-industrial agrarian societies, and thus the development of the new organizational form of the construction industry was dependent on the emergence of the social technology of contracts, free labor markets, etc. Furthermore, Stinchcombe argues that there is a correlation between the age of an organizational form and its characteristics.

The second part of Stinchcombe's hypothesis accounts for the stability and longevity of organizational forms once they are established. According to him, the construction industry of today has largely the same organizational form as it did at its inception in pre-industrial cities of Europe.[11] This form has persisted despite many attempts to 'rationalize' or 'industrialize' the construction industry.

Stinchcombe's analysis of the relations between social structure and the development of new organizational forms is very insightful. However, although he discusses the importance of 'technical and economic conditions', he does not stress the role of technology in the development of new organizational forms. To be fair, Stinchcombe is primarily concerned with the influence of social structure and not technology, and therefore treats the latter as a set of conditions in a manner similar to that described above. Nevertheless, the attributes of technology, rational problem solving, and division of labor, do seem to influence how a new organizational form is developed. In fact, significant changes in these attributes are a measure of technological change.

Alfred Chandler takes up the argument for the development of new organizational forms where Stinchcombe leaves off and explicitly assigns a role to technology.[12] He postulates that technological advances in the methods of production increased the rate of production to the point where new organizational forms were necessary to manage the enterprises. If Chandler's perspective is added to Stinchcombe's the following proposition can be offered: when new technologies are developed under supportive social structural conditions, new forms of organizations are invented to develop and implement the new technologies. If successful, these new forms embody and promote technological change. They also drive older, less competitive organizational forms out of business. Furthermore, these forms remain stable, and account for the consistency of technological practice within industries we see today.

At this stage of research, the evaluation of the technological change – organizational form hypothesis can be discussed only in terms of a series of cases which appear to fit the pattern. These cases include the machine tool industry, steel manufacture, automobile manufacture, solid state electronics companies, and emergency medical care (paramedical) care organizations.

(a) The machine tool industry first developed in England in the second half of the eighteenth century, and was transplanted to the United States in the early decades of the nineteenth century.[13] One attribute of this organizational form is small size. The machine tool industry represents an organizational form that exhibits little growth. The 1977 Census of Manufacturers reports the following size distribution: 1 to 9 employees, 54%; 10 to 49, 28%; 50-249, 12%; and 250 or more, 6%.[14] Another attribute of the industry is the way different branches are linked together in a social network. New firms are often spinoffs from established firms. Moreover, firms are very specialized, each occupying a niche defined by product and customer(s). Thus, direct competition is reduced and the industry network also serves to channel customers to the appropriate shop. A third important attribute of the machine tool industry is the apprenticeship mode of training. During the latter part of the eighteenth and early nineteenth centuries, this was the primary means of gaining a technological education. While today engineers study machine design, in general, in engineering schools, the machine tool industry still practices the art of technology as much, if not more, than the science of technology in the development and production of new machine tools.

(b) In the 1850's, the first large scale integrated iron works, the American Iron Works, was built in Pittsburgh.[15] That mill became the Jones and Laughlin Steel Company. An engineer named Laughlin figured out how to integrate several major steps in metal manufacture that were previously carried out by small, inefficient, and separate companies. At the time of this development 'large' companies employed about two hundred people. Within a relatively short period of time, the J&L Works employed two thousand people, an indication of the increase of scale associated with the innovation. Moreover, this plant set the pattern for all the mills; it appears that they either adopted the new organizational form, or ceased to operate. Finally, contemporary mills are still organizationally similar to nineteenth-century mills.

(c) In 1914, Henry Ford introduced a new organizational form of great significance when the River Rouge plant opened in the Down River area of Detroit.[16] While Ford had been producing the Model T for several years, he was unable to achieve major economies of scale necessary to capture a mass consumer market until he developed and implemented his new assembly technology. He was able to combine the standardized parts of the American system of manufacturing with the moving assembly line. Most important, Ford developed the modern industrial bureaucracy to manage the whole production process. His bureaucracy was so complete he even had a Sociology Department to improve the morale and morals of his employees. Again, other automobile companies rapidly adopted Ford's technology *and* form of organization, or they went out of business. Most went out of business. The Ford Motor Company's success was almost its downfall. By 1920 *all* the surviving automobile companies had grown to the point where they were unmanageable with Ford's form of industrial bureaucracy. This situation led to yet another organizational form.

(d) In 1920, Alfred P. Sloan of General Motors invented what has come to be known as the modern corporation.[17] He developed a system that delegated operating decisions to divisions, and centralized financial controls and long range planning into a newly formed corporate staff structure. General Motors Research Laboratories were also established as a corporate function and the activities were integrated into the corporate planning process. While this social invention was not tied directly to a new technological development, it made possible a tremendous increase in the *scale* of the automotive technology, and it also institutionalized, through the corporate planning function, the regular technological change of the product and the manufacturing facilities. While Henry Ford had not yet tackled the problem of what should come after the Model T, General Motors already had institutionalized, market-oriented technological change. While General Motors is many times larger today, its organizational form still fits the design laid down by Sloan in 1920.

(e) Companies and industries that have developed since World War II are generally less bureaucratic than the General Motors type of corporation. The matrix organization and the extensive use of engineering task groups are more common in organizations formed or reformed in the last forty years. Most of the companies in electronics and computers fit this form, with IBM, Hewlett-Packard, and Texas Instruments being prime examples.[18] Planned technological change is also important in influencing the organizational forms of these electronic firms. The stately pace of technological change in the automobile and other consumer durable industries is too slow to survive in the modern electronics market. Moreover, electronics companies are less concerned with the economies of scale in manufacturing. Their main concern is to maintain a high rate of technological improvement in their products. To do this, they have organized around the product development cycle. Thus matrix structures facilitate management of the essentially horizontal development cycle. Another important feature of the organizational form of electronics firms is the relatively small size of their operating units and divisions. Because it is more difficult to bring about technological and organizational change in a large organization, many firms keep their divisions small, deliberately sacrificing economies of scale in production to maintain flexibility in technological development. Two prime examples are Hewlett-Packard and Texas Instruments. While both are large companies, they try to keep their divisions small. Hewlett-Packard splits up divisions that grow much above twelve hundred employees, and Texas Instruments has over one hundred divisions.

(f) The final case is the new form of ambulance service that has evolved during the last decade. The origins of this new organizational form are in the military medical experience in Viet Nam. The military developed both the technology and the organization to substantially improve survival chances of wounded soldiers. During the latter part of the 1960's, it was common to read in the newspapers that the soldier wounded in Viet Nam had a better chance of surviving than someone involved in an accident on a Los Angeles freeway.

The medical care innovation was to take medical care to the person at the scene of injury and begin treatment at once. To do this required a completely different ambulance vehicle and people with medical training relevant to emergency situations. The new organizational form was actively promoted by the Federal government and now most ambulance services in the United States are of this type.

6. COMMON OBSERVATIONS

These examples share a number of interesting features that I shall now list:

First, a single technologically and organizationally innovative company often set the standard for the whole industry. In many instances, those companies are still the industry leaders. In the ambulance case, the Army Medical Corps served as a model.

Second, significant changes of scale were often associated with these organizational and technological innovations. These scale changes had major economic impacts.

Third, in several cases, the organizational mix of the art and science of technology appears to have shifted. For example, Ford's introduction of the moving assembly line and extreme division of labor removed most of the art of technology from the shop floor. Sloan's decentralization of operating decisions allowed greater opportunity for plant managers and engineers to exercise their know-how. The development of the new emergency medical care ambulance service involved a significant increase in the science of technology.

Fourth, among the more modern industrial cases, organizational innovation generally includes the institutionalization of technological change. Product development cycles became standard operating procedures. Most recently, companies have been adopting organizational forms that attempt to continually create a high rate of technological change. It remains to be seen whether these organizations can resist the historical pattern of slower growth and lower rates of change.

Finally, returning to the theoretical arguments of Stinchcombe and Chandler, two factors seem significant in most, if not all, of these cases. First, technological change increased productivity and economic competitiveness to such an extent that older forms of organization were unable to successfully manage the new technologies. Thus demand for a new form is created. Second, these changes and the advantages they created were great enough to overcome what Stinchcombe calls the 'liability of newness'. That is, significant technological advantage is able to subsidize the organizational learning process necessary to create a new organizational form. Such organizational development is difficult, costly, and can result in very inefficient operations.

Department of Sociology
University of Pittsburgh

NOTES AND REFERENCES

1. Among the more commonly cited references that discuss organizations and technology are: James D. Thompson, *Organizations in Action* (New York: McGraw Hill, 1967); Joan Woodward, *Industrial Organization: Theory and Practice* (Oxford: Oxford University Press, 1965); Charles Perrow, *Complex Organizations: A Critical Essay* (Glenview, Illinois: Scott, Foresman, 1972); and Robert Blauner, *Alienation and Freedom* (Chicago: University of Chicago Press, 1964).
2. Thompson, *Organizations in Action*.
3. Norman P. Hummon and H. E. Hoelscher, 'Technological Change as a Societal Process'. TECHNOS, July-September, 1977.
4. This example is based on an interview with a design engineer with the Boeing Corporation.
5. Norman P. Hummon, Reginald Baker, and Linda Zemotel, 'Survey of International Technology Transfer From the United States: The Viewpoint of U.S. Technology Suppliers.' Institute for Research on Interactions on Technology and Society, University Center for International Studies, University of Pittsburgh, 1978.
6. This example is distilled from numerous discussions with a chemist who worked in the development laboratory of a major chemical company, and from talking with H. E. Hoelscher, Professor of Chemical Engineering and former Dean of Engineering, University of Pittsburgh. Dr. Hoelscher has consulted with several major chemical companies.
7. Alan H. Cottrell, states that "Traditionally it has taken some 20 to 50 years for a scientific advance to work its way through to full commercial exploitation, but under the pressure of technological competition this time has recently contracted considerably and is now sometimes no more than five to ten years," in 'Technological Thresholds,' in *The Process of Technological Innovation* (National Academy of Engineering, April 24, 1968), 50.
8. Paul R. Lawrence and Jay W. Lorsch, *Organization and Environment* (Homewood, Illinois: Richard D. Irwin, Inc, 1969).

9. Lawrence J. White, *The Automobile Industry Since 1945* (Cambridge, Mass.: Harvard University Press, 1973).
10. Arthur Stinchcombe, 'Social Structure and Organizations', in James G. March, ed., *Handbook of Organizations* (Chicago: Rand McNally, 1965).
11. Arthur Stinchcombe, 'Bureaucratic and Craft Administration of Production,' *Administrative Science Quarterly* 4 (1959).
12. Alfred Chandler, *The Visible Hand: The Managerial Revolution in American Business* (Cambridge, Mass: Harvard University Press, 1977).
13. Eugene S. Ferguson, 'Metallurgical and Machine-Tool Developments' in Melvin Kranzberg and C. Pursell eds., *Technology in Western Civilization* (Oxford: Oxford University Press, 1967), and Edwin Layton, 'Technology as Knowledge,' in Layton, ed., *Technology and Social Change in America* (New York: Harper and Row, 1973).
14. Census of Manufacturers, Bureau of the Census, 1977.
15. This discussion of the organizational form of the iron and steel industry is based on research by Dr. Reginald Baker, formerly of the University of Pittsburgh. An 1869 *Scientific American* and historical census materials for Pittsburgh were used in the research.
16. Several good discussions are available of Henry Ford's technological and organizational innovations. See James J. Flink, *The Car Culture* (Cambridge, Mass.: MIT Press, 1975); John B. Rae, *American Automobile Manufacturers: The First Forty Years* (Philadelphia: Chilton, 1959) and Reynold M. Rik, *Henry Ford and Grass-Roots America* (Ann Arbor, Mich.: University of Michigan Press, 1972).
17. Alfred P. Sloan Jr., *My Years with General Motors* (Garden City, New York: Doubleday, 1964), and Alfred Chandler, *Strategy and Structure* (Cambridge, Mass.: MIT Press, 1962).
18. Thomas J. Peters, 'Putting Excellence Into Management,' *Business Week* (July 21, 1980).

Rachel Laudan

COGNITIVE CHANGE IN TECHNOLOGY AND SCIENCE

1. INTRODUCTION

The central presupposition of this paper is that cognitive change in technology is the result of the purposeful problem-solving activities of the members of relatively small communities of practitioners, just as cognitive change in science is the product of the problem-solving activities of the members of scientific communities. If this is correct, we ought to be able to develop theories of cognitive change in technology analogous to the theories of scientific change that have been advanced in the last couple of decades. My purpose in this essay is to explore this possibility.

In doing so, I shall set on one side many aspects of technological change that are of central concern to other students of technology, ignoring, for example, the social and economic causes of such change, as well as the effects of technology in a wider social setting. I have given my rationale for restricting the meaning of technological change in this way and for these purposes elsewhere.[1] Furthermore, unlike many of the other authors in this volume, I shall concentrate more on the mechanisms for technological change than on the nature of the technological community. In thinking through this paper I have necessarily drawn extensively on recent work in the theory of scientific change, particularly that by Thomas Kuhn, Imre Lakatos, Larry Laudan and Thomas Nickles.[2] I shall not cite their work line by line since I am not applying the *specific* doctrines that one or other of these authors espouses to the study of technology. Given the present stage of development of models of technological knowledge change, even the most general categories still have to be carved out. Thus my strategy will be to identify some of the more widely shared taxonomic categories in the theory of scientific change and pursue their applicability to cognitive change in technology. One further introductory remark is in order. Since my primary concern is the internal intellectual development of technology I shall not

R. Laudan (ed.), The Nature of Technological Knowledge, 83–104.

seek, except incidentally, to illuminate the relationships *between* science and technology. While a knowledge of recent work in philosophy of science *informs* my analysis of technology it is technology that is the subject of this paper *not* the relationship between technology and science.

2. TECHNOLOGY AS PROBLEM-SOLVING

To say that problem-solving constitutes the major cognitive activity of the technological practitioner is scarcely new.[3] Change and progress in technology is achieved by the selection and solution of technological problems, followed by choice between rival solutions. Viewed this way, cognitive change in technology is a special case of problem solving. Yet few scholars have yet attempted to compare and contrast technological problem-solving to problem-solving in other areas of human activity. Since most models of scientific change developed in the last two decades see problem-solving as characteristic of science, the way is clear to explore analogies between scientific change and technological change.

I should, in passing, deflect one potential criticism of this approach to technological change. It is sometimes claimed that to consider technology as no more than problem-solving is to neglect the aesthetic drive, the cultural concerns and the sheer fun and enjoyment that motivate many technologists, and to focus only on those rather dreary utilitarian elements that the wider public tends at present to identify with technology. In the sense in which I understand problem-solving this concern disappears, since for me technological problems include not only the utilitarian puzzles but also the aesthetic and intellectual issues that engage the practitioner. To restrict technology entirely to mundane issues is to ignore the historical record which establishes beyond a doubt the wider scope of technology.

An appropriate first step in understanding technological change is to attempt to construct a taxonomy of the different kinds of problems that technologists face.[4] The first type of problem, and one occurs very rarely, is a problem given directly by the environment and not yet solved by any technology. Examples would be the flooding of a village on a regular basis, or the deterioration of food stuffs when stored. A clever practitioner may perceive these as problems (rather than as the inevitable course of events), and moreover as problems that are *technically* soluble, and go on to invent the first dam or the first clay pot. Although this fairy tale captures much of the popular image of the inventor, in fact such immediate perception of difficulties rarely provokes a technological response *unless there is already an extant technology that is directly applicable to the situation, or that can be suitably modified.* Instead, the reaction to such hazards tends to be fatalism or a social or economic solution. The villagers move away from the river or resort to infanticide to reduce their numbers. Put another way, the problems that technologists tackle only occasionally come from a perception of the world unmediated by technology. Social 'fixes' usually come to mind more naturally than technological 'fixes'.[5] This is not *always* the case, or we would still be hunting with stone axes, but as time has passed we have come to live in an increasingly technological environment with a widening range of skills at our disposal, and we find the chief locus of technological change in modification of extant technology rather than in invention in direct response to environmental hazards.[6] Indeed the very difficulty of thinking of contemporary problems posed by the environment and perceived as technological problems in the absence of appropriate technological traditions bears testimony to this point. All the remaining categories of technological problems arise out of extant technology.

A second source of technological problems is the functional failure of current technologies.[7] Such functional failure can occur when a technology is subject to ever greater demands or when it is applied in new situations. Bridge failure, to take an obvious example, usually occurs when loads are higher than usual. In other cases, functional failure may only become obvious after the technology has been in use for some considerable time. The inadequacy of canals as the major British transport system was only recognized once the scarcity of water on a relatively small island became obvious. But the assumption sometimes encountered that functional failure is an insignificant source of technological problems, on the grounds that no technology that failed to work would be implemented, is misleading. Technologies are implemented as much because they are needed as because they work successfully. Need and feasibility are not co-variant and history is strewn with examples of technologies that were adopted because there was such a crying need for them, not because they worked. Medical technologies, at least until the twentieth century, fall almost entirely within this class.

A third way in which problems are generated is by extrapolation from past technological successes. I shall call these cumulative improvement problems. The mechanician that has successfully built a spinning machine with 50 spindles is likely to see the production of a similar machine with 100 spindles as his next problem even though there is no failure in the machines he has already built. Similarly an aircraft industry that has increased the speed of planes from 100 to 300 miles per hour perceives its next task as achieving speeds of 350 miles per hour. The internal dynamic of technology itself, isolated from more widespread economic considerations, is a potent source of technological problems. Many problems of intellectual, rather than practical importance are generated this way, and tackled by the technologist for the sheer excitement of determining whether they are soluble.

Fourth, and very important, imbalances between related technologies in a given period are often perceived as technological problems. These imbalances, as both economists and historians have pointed out, are fertile problem generators. The practitioner, surveying the state of technology, notices that the effective operation of a particular technology is being impeded by the lack of an adequate complementary technology. Particular variants of these situations have been variously called reverse salients,[8] backward linkages,[9] compulsive sequences,[10] and technological co-evolution.[11] The imbalance may be primarily economic, as in the English textile industry in the eighteenth century, or it may be technological, as in the case of the stability problem in ocean-going ships after the change from wood and sail to iron and steam. In either case the solution can be technological.

Finally, some problems are perceived as potential rather than actual failures in a technology, their future occurrence predicted by some other system of knowledge, most notably science. Constant has claimed that such 'presumptive anomalies', were the chief stimuli provoking the turbojet revolution, since aerodynamic theory in the late 1920s predicted that piston-propeller planes would fail if higher speeds and altitudes were attempted.[12] This led those engineers who were sensitive to aerodynamic theory to develop alternatives before piston-propeller planes had failed, and indeed while they were still undergoing steady improvement.

3. THE SELECTION OF TECHNOLOGICAL PROBLEMS

There are always more problems awaiting solution than can be realistically tackled at any one time. The practitioner does not select problems at random from this pool but chooses them in light of his overall goals. The literature on technology normally stresses the central importance of social and economic concerns in the selection of problems for further investigation, and without doubt this analysis is often entirely correct. In many instances the technologist assigns importance to problems in accord with the social or economic utility assigned to them by society at large. But, in numerous other cases problem importance is correlated with centrality within technological systems. If we have an entire edifice of related technologies ready to go and only one technology is missing, then we shall assign great importance to solving that problem. Needless to say there is often a substantial overlap between problems that are weighted high because of their social importance and those that are important missing links within technological systems.

Yet there is another factor at work in the selection of problems which is worth emphasizing. In particular the technologist is likely to select problems that he believes to be soluble. It is a cliché of the huge literature on problem-solving that we naturally tend to shy away from problems that we have no notion how to solve. Thus practitioners are unlikely to select as targets those problems, however socially and economically pressing, that show no sign of being soluble. Perceived solubility acts as a filter, and problems that do not pass it tend to be ignored, however much other criteria might indicate that they should be selected. How do we judge which problems are soluble? The answer is basically in light of our past experience. Much of the work on problem-solving strategies in the cognitive psychology literature boils down to this homespun wisdom. Similarly work in the history and philosophy of science over the past few decades has shown that the scientist does not normally approach the natural world directly, but rather brings to his investigations a familiarity with the science of the day. This background knowledge in many cases determines, and in most cases influences, the subjects he seeks to investigate; frequently it also shapes the particular form of the investigation. Thus the scientist works within a matrix of contemporary science, building on it, and reacting to it. Similarly most technological innovation takes place within a matrix of current technology. The technologist does not begin to construct technologies from scratch, but works in a well-structured world of technological knowledge, which enables him to select problems that he believes to be soluble. The major exceptions to this are presumptive anomalies, of which more later.

The cognitive structure of the technological world has two main dimensions, one historical and one structural, both of which are familiar to technologist. He is, for example, cognizant of the main historical traditions, at least in the immediate past, in his area of technology. Whether he be a millwright or an aeronautical engineer, he will be familiar with the range of approaches used by his immediate predecessors in the field. Similarly he will be aware that the technology of his own day is more than a collection of isolated inventions and processes, and that it is arranged in systematic structures. In the same way that ideally the different branches of science are at least consistent, and better still well-integrated, so too is much technology organized systemically. The reasons for this may be somewhat different in the two cases; inconsistent scientific theories leave us unsure what to believe; ill-integrated technologies simply fail to work. But whatever the reason for the structured nature of scientific and technological knowledge, the practical result is the same – knowledge not only of the specific problem but of its context affects the practitioner's selection and solution procedures.[13] The practitioner therefore has to operate with an understanding, albeit incomplete, of the technological traditions leading up to his present situation and the technological systems currently dominant in his era and area. When not fully conversant with this background at the start of his work, the practitioner frequently goes to tremendous trouble to acquire it.[14] Without such knowledge, both about problems and about possible ways of solving them, he is hard put to produce successful new technology, a situation which explains why so much invention, at least of a day-to-day sort, is 'in-house' in the R&D lab of the corporation, or, in earlier times, within the firm or family.

4. THE SYNCHRONIC STRUCTURE OF TECHNOLOGICAL KNOWLEDGE

For purposes of understanding the cognitive situation of the practitioner, three levels of technological knowledge, with their associated artifacts or processes, can be distinguished at any one time. I shall call these levels individual technologies, complexes and systems.[15] Obviously these categories are distinguished for analytic convenience, and there is considerable overlap between them. Nonetheless I believe that they enable us to analyze the mechanisms of technological change more precisely.

The lowest level, that of the individual technology, can be usefully thought of as analogous to the level of individual theory in science. Just as theories are solutions to scientific problems, so technologies are solutions to technological problems. They also correspond to what in common parlance is called an 'invention' though I shall not use this term as it carries with it too much freight in its suggestion of isolation from extant technological knowledge and serendipitous discovery. I want to stress by contrast that most technologies (or inventions) are the result of systematic and rational problem solving rather than pure luck. Examples of individual technologies would be the Newcomen engine, the Britannia Bridge, or the gyroscope. The practitioner's environment is made up of numbers of such individual technologies, and a large proportion of his energies are expended at this level. The problems he selects, the methods he uses to solve them, and the assessment of solutions is frequently carried out with respect to individual technologies. Many of the problems the practitioner tackles, particularly functional failures, cumulative improvement problems and presumptive anomalies occur at this level. Even those problems that are perceived as imbalances higher in the technological system are soluble primarily by new individual technologies. The heuristic that the practitioner employs in designing individual technologies is in most cases derived from past successes in this level of problem solving. Finally the adequacy of an individual technology will be assessed in relationship to other individual technologies. Thus for problem generation, problem solution, and assessment, the individual technology is the focus of much technological change.

Not surprisingly individual technologies tend to be relatively short-lived, at least when compared with complexes and systems. There are exceptions, as always, to this generalization especially where simple technologies are concerned. The safety pin, to take but one example, has persisted relatively unchanged since antiquity. More frequently however new technologies that out-perform their older rivals in a given environment tend to replace them at regular intervals. As we shall see shortly, many of these individual replacements occur not at random, but in coherent sequences which I shall call traditions.

Yet most individual technologies serve little purpose on their own. The Newcomen engine was idle without mines from which to pump water; the Britannia Bridge was no more than a stunning feat of engineering without a railroad network,[16] and a gyroscope remains a toy if not integrated with other technologies. We find that most individual technologies are linked together in *complexes*, the second level of technological structure roughly analogous to that of 'super-theory' in science. Technological complexes can stand in relative isolation in the production of a good or service. A cotton mill can produce finished cloth from raw material in one location. A canal system can transport materials from Lake Erie to the Hudson. Unlike individual technologies that rarely deliver a finished good or product, complexes frequently do. Compared to the individual technologies that constitute them, these complexes tend to be long-lived. In large measure this is due to the fact that individual technologies within the complex may change but provided that the *relationships* between the technologies making up the complex remains the same, we may say that the complex has not changed. Complexes in most cases can outlast changes in the individual technologies that constitute them. The Cornish beam engine could replace the Newcomen engine in many mines. Truss bridges could be substituted for tubular steel bridges of the Britannia Bridge type in many railroad systems. This is not to deny that in certain instances the introduction of a new technology to an existing complex can precipitate a radical overhaul of the whole complex. The appearance of turbojets, for instance, put such strains on airports designed for smaller, slower piston-propellor planes that airports had to be re-designed in order to handle the new technology.

The complex, although normally much more complicated than the individual technology, is still a level that the practitioner can grasp as a whole. Thus the technologist can discern imbalances in complexes, whether they be between the spinning and weaving phases of the textile industry in the eighteenth century, or the inadequate braking systems on locomotives in the nineteenth. Many of these imbalances can be dealt with by social and economic means: wages can be adjusted, laws can be passed in order to keep the complex working smoothly. What distinguishes the technologist from the businessman, the legislator or the ordinary member of the public is that he attempts to correct these imbalances technologically. Rather than adjusting wages or lobbying to get legislation passed, he invents mules, throstles, power looms and air brakes. Thus he solves imbalance problems by developing solutions at the level of the new individual technology.

Very occasionally, an entirely new complex is introduced wholesale at an identifiable point in time, rivalling, if not immediately displacing, an older complex. A classic example is Edison's systematic and conscious challenge to the gas lighting complex with his introduction of an electric lighting system.[17] Schumpeter coined the terms 'innovation' to identify the event, and 'entrepreneur' to describe its architect.[18] Entrepreneurs need technical skills, but in order to design systems on this scale they have to employ many other skills as well, economic, social and political. Typically, although they may invent one or two components of a new system, much of their expertise lies in re-combining extant technologies into new complexes. This type of change is analogous to that called a revolution in science, and is the kind of event that Constant was pointing to in talking about the turbojet 'revolution'. I shall discuss these unusual events in a little more detail at the end of the paper.

The degree of integration of a complex, and the extent of its imbalances, has an important effect on the rate of technological change. In order for practitioners to generate technological change at a regular rate, neither destructively fast nor crampingly slow, the complex needs to have a *fair*, but not complete degree of integration. With a highly-integrated complex, stasis rather than technological change becomes the more common phenomenon since few imbalance problems can be perceived by the practitioner.

Technological complexes in the ancient and medieval periods tended to be rather rudimentary, with relatively few close linkages. Integration was easily achieved, and the pressure for change was minimal. In the post-Industrial Revolution period, the linkages are much tighter and much more frequent. Complexes become much harder to integrate, since, as we have seen, the individual technologies that are the component parts of any particular complex have a tendency to change. A productive technological society must strike a balance between well-integrated smoothly working complexes and lack of integration leading to technological change. Too little integration and the efficiency, productivity and usefulness of the complex is impaired. Too tight an integration and an important stimulus for change is absent.

There is a third yet higher level of technological organization, namely the system. This corresponds roughly to the level in science that has sometimes been called a scientific world view. Wide-ranging in space as well as long-lived in time, attempts to identify systems tend to produce rather nebulous results. But just as the history of science can, for convenience and to a rough approximation, be divided into the Aristotelian, Newtonian and Einsteinian world views, the history of technology can also be categorized in terms of epochs gradually replacing one another. The system of wood, wind and water has been replaced by a system of iron, coal and steam, and this in turn displaced by electricity and synthetics. Several scholars have attempted to periodize the history of technology, including Mumford (eotechnic, paleotechnic and neotechnic) and Gille (the Egyptian-Mesopotamian, the Graeco-Roman, the Medieval, the Renaissance-Classical, the Industrial Revolution, the Modern and the Contemporary).[19] Assuming that historians are correct in assuming that something like these systems do exist, they, like complexes, have to be integrated. Lack of integration between different complexes within an overall system has been identified by many authors as a source of technological change. Nonetheless technologists rarely perceive systemic problems as a whole. Eighteenth century practitioners did not say to themselves "wood and water are running out, and we must find an alternative technological system." Instead they perceived the problem at the level of the individual technologies involved, or, at the highest, at the level of complexes. They tried to replace charcoal by coke, wood by iron *in specific circumstances*. There is no analog for the entrepreneur who replaces one complex with another when one moves to the level of systems. When one system replaces another it is a piecemeal and long drawn out process. Thus although it is important to identify these systems for our understanding of the sweep of the historical process, they are not so essential for understanding the mechanisms of technological change, the cut and thrust of the individual actions and community values and sanctions that bring about shifts in technology.

There are some features that systems, complexes and technologies all share. For example, in one sense they are *all* systematic. Individual technologies, complexes and systems are all made up of a number of parts such that the failure of any one of those parts affects the performance of the whole. Only the very simplest technologies are not made up of inter-connected parts – the stone tool or the digging stick for instance. Anything more sophisticated is likely to be systematic, but this need not lead us to analytic despair. As Herbert Simon and, following him, Edward Constant have pointed out, these systems are arranged hierarchically.[20] For example the fire and the boiler forms a subsystem within a steam engine; a steam engine may be a subsystem within a textile mill; and an individual textile mill is part of the whole process of producing textiles. These systems in Simon's terms are *decomposable*. That is to say, we can treat individual subsystems as units because they can be replaced by a similar or different subsystem without affecting the whole. For example, boiler design changed dramatically between 1780 and 1850, without changing the role of the steam engine in providing rotary power. Thus we can treat systems at a certain level of generality as our subject, without worrying overmuch about the subsystems from which they are constructed or the super-systems of which they are a part.

5. THE HISTORICAL DIMENSION OF TECHNOLOGY

Traditions are the historical component of the technologist's cognitive situation. They are easy enough for the historian to identify: for example, there is the clock and instrument-making tradition, the machine tool tradition, the water turbine tradition and so on. Definitions of traditions are somewhat harder to come by and probably misleading in any case, though each tradition can be roughly characterized as consisting of a community of practitioners applying evolving methods to solve sequences of problems. Often the relationship between successive problems and solutions within a tradition is very obvious, as in the case of successive designs of water wheels for increased efficiency under varying conditions. Sometimes the relationship is more tenuous, as when the turbine tradition was extended from water to air. Traditions evolve either by the application of a technique to a new situation, or by improving that technique in the same situation. It is important to notice that technological traditions occur at the level of individual technologies. Though developing sequences of complexes are sometimes found, the lines of continuous development in technology are generally traced through individual technologies. Progress in complexes and systems occur by alternations of imbalances and improvements in the technologies making up the larger unit.

Sociologically there is a great deal of continuity within traditions. Technological knowledge is passed down from father to son, from master to apprentice, from engineer to student. Every technological tradition has a shifting but sustained community of practitioners associated with it. The importance of this community in technology transfer is well-established. Recently scholars like Merrit Roe Smith and Anthony Wallace have begun to demonstrate how necessary sustained and appropriate structures are for the continuity of the community of practitioners over a number of generations.[21]

Understood this way, the community of technologists with which I am concerned is the community of technology generators. There is an alternative way to divide up the technological community. One could stipulate that a community be defined as that collection of individuals who generate *and produce* the technology. This would mean the inclusion of technicians, business men, salesmen and so on. Much confusion has been introduced into the discussion of technological change by including within the community anyone remotely connected with that change. Communities do overlap, and individuals have places in more than one community. Thus we can quite legitimately talk about a community of technologists as the generators of technology, without for one moment denying that there are many other people in society, from laborers to financiers, who are involved in other kinds of technological community. We do not feel constrained to include laboratory technicians and high school teachers in the community of resarch scientists, even though both have a crucial role to play in a science-oriented society. We should feel free to make a similar distinction where technology is concerned.

The identity of a tradition rests in this continuous historical sequence. The tradition may well develop in such a way that its starting point and its end point, when viewed individually, have no obvious similarities. There is scant resemblance between the sophisticated turbines of the late nineteenth century and the simple horizontal water wheels from which they are descended, nor are the original technologists in the community still alive. Only when all the intermediate steps are traced can the tradition be detected. I am therefore not identifying traditions by insisting that one or more of their attributes persist throughout the life of the tradition. Nor am I arguing that a tradition can be defined throughout its development by the employment of a particular method, or one static problem, or the use of a characteristic material. Often methods and problem sets associated with a given tradition will persist through time, but, provided there is historical continuity, I am prepared to allow dramatic modifications in one or even in all of them. Conversely, I do not count as belonging to the same tradition sequences of techniques independently developed in different parts of the world, or lost and rediscovered. On my account, methods of timekeeping and calendar-making dependent on megaliths and monuments which were developed independently in different parts of the world do not constitute a single tradition. Nor does it seem particularly fruitful to me to begin analysis of steam engine development with Hero of Alexandria since his work appears to have had no significant impact on the experimenters of the seventeenth and eighteenth centuries.

One very important function of traditions is to focus the technologist's attention on potentially soluble problems, particularly those of cumulative improvement. The tradition suggests both the possibility of improving the performance of some technology and the methods to achieve this improvement. It is probably true to say that the bulk of the work done in the development of technology is on cumulative improvement problems. But the problem-generating function of traditions is inextricably linked up with two other functions, the heuristic and the assessment functions, to which I shall now turn.

6. PROBLEM-SOLVING HEURISTICS

The heuristic associated with traditions often indicates what steps should be taken to improve the technology, just as the heuristic associated with scientific traditions often indicates the way in which theories should be changed. In Ptolemaic astronomy, for example, unexpected perturbations in the movements of the heavens could be accommodated by including another epicycle or equant in the theoretical model. Similarly, a clockmaker in the Renaissance who wished to replicate even more of the celestial motions on the face of his clock could do so by adding further gear trains to the already complex motion. Of course as in science, the heuristic is more effective in some traditions than others. Large scale visible mechanical systems lend themselves to improvements in a way that systems dealing with the structure of matter do not. In metallurgy, for example, the heuristic may well amount to no more than the commonsense, trial and error recommendation 'try another additive'. Indeed it is precisely in these kinds of technology that science, when advanced enough, can come to play a significant role. Clearly the problem-generating and heuristic aspects of a tradition are two sides of a single coin. No one likes to tackle problems he believes himself unable to solve. Problems generated within a tradition simultaneously offer hope of solution by the further application of methods already developed within that tradition. However tradition-based heuristics do not provide the only methods of approaching technical problems.

A second source of strategy that is very commonly used in tackling technological problems is to apply expertise developed in one complex to problems in another complex. Thus in the nineteenth century skills acquired in machine tool design were applied in a wide variety of new situations including small arms manufacture, sewing machines, typewriters and bicycles to name but some of the more prominent. This is the phenomenon that has been described as technological convergence.[22] In a similar way, an inventor-entrepreneur like Elmer Sperry selected those problems where his knowledge of electrical technology offered promise of successful application in a new area.

A third fruitful source of heuristic aid, at least in the past hundred years, has come from outside technology, most particularly from science. Although scientific knowledge usually cannot be translated directly into technological expertise, it can offer the technologist a fresh perspective on old problems. The heuristic aid that scientific knowledge can offer the technologist is in part what Constant had in mind when he developed the notion of presumptive anomaly. However I believe some careful distinctions have to be made here. Presumptive anomalies, broadly understood, can be divided into two classes: first there are those where a knowledge system outside technology gives reason to believe that a technology will fail; and second there are those where, in addition to the predicted failure, the system that predicts the failure simultaneously offers hope of a solution that "will do a much better job."[23] This latter case is, I believe, the one that Constant was referring to in the case of the turbojet revolution, and explains part of his resistance to extending the notion of presumptive anomaly beyond science. Nonetheless his own definition includes both classes. In the first very common but weak sense of the term 'presumptive anomaly', economic and demographic forecasts of energy reserves and population growth, for example, have prompted the development of new technologies, although they patently give no heuristic guide to the technologist in constructing those technologies (except insofar as they constrain the range of acceptable solutions). Even where the *physical* sciences rather than the social sciences predict failures, it is far from clear that this prediction always carries with it hints for alternative solutions. Indeed one suspects this is the exception rather than the rule, so that in general we just take such predictions as indications of our human limitations. Space science and technology is replete with instances where science predicts that we will not be able indefinitely to go faster or further, but where no hints of a new technology to overcome this block are forthcoming. Before being widely applied to the history of technology, the term 'presumptive anomaly' needs much more careful analysis and definition than it has thus far received.

7. ASSESSMENT OF NEW TECHNOLOGIES

A further important activity for the technological community, in addition to problem selection and solution, is the assessment of those solutions. The primary assessment of new technological solutions is carried out by the practitioner and his immediate community. Only they are in a position to know whether or not the new solution counts as an improvement over previous solution, or to assess if the new solution's performance is better than the previous technology in certain respects and worse in others. This fact about the primacy of assessment of new technological solutions by the practitioners themselves is often obscured because other communities frequently hold veto power over further development and implementation of the technology. Military leaders, business men, or in some cases, the general public all have the ability to reject solutions acceptable to the technological community. Their criteria may well differ in kind and degree from the ones employed by the technologists themselves, and I shall not discuss them further here. Technologists bring a number of different criteria to bear in deciding what constitutes a satisfactory solution to a technological problem. The ones that first spring to mind are feasibility and cost. Further reflection suggests that at least for certain technologies (some kinds of building for example) aesthetic criteria are also important, and that in some cases cost is a relatively insignificant factor (military technology). There is an interesting story to be written about the selection and weighting of assessment criteria for new technologies. This is not the place to tell that story. The point that I want to stress here is that the assessment is always comparative. That is to say, technologies are not judged in the abstract with respect to whether they work, how much they cost and so on, but in relation to other technologies. Thus a new design of water wheel is judged with respect to its ability to deliver power for a given volume and fall of water when compared to other water wheels. The choice is comparative even in the absence of rival technologies, since we can always choose between the proposed technology and *no* technology.

Since choice and assessment are comparative, traditions have a further important role to play. With respect to closely similar technologies succeeding one another in a tradition, relative judgements are comparatively straightforward since all the assessment criteria (cost, strength, reliability, efficiency etc) can be held constant. Working within a tradition the practitioner can usually decide about whether or not his solutions to the problems presented by that tradition and solved in terms of its heuristic do or do not constitute improvements over previous solutions. When there is an abrupt change of tradition this becomes much more difficult. To give a specific example, once water shortages made the relative efficiency of different designs of water wheel of more than merely academic concern John Smeaton was fairly readily able to develop ways of assessing efficiency for these machines. However when steam engines began to replace water wheels as power generators, comparative assessments became much more difficult. Boulton, Watt and others had to put an enormous amount of effort into producing a new measure, horsepower, for just one criterion of assessment, power output.

Put another way, in any given technological tradition standards of assessment tend to be fairly easy to establish. Users may query *which* attribute is most important (economy or durability, say) but they will not have much difficulty comparing technologies with respect to each attribute. The larger the water wheel, the greater the power generated. When a revolution occurs (water to steam) it may well be hard to find a common yardstick for the same attribute. Furthermore, the weighting of the various attributes may be very different before and after the revolution. Such revolutions thus become an important stimulus to the generalization of technological thought, to seeing problems in a more synthetic and less particular way, and to the development of explicit methods of testing technologies.

8. COMPETITION AND CO-EXISTENCE

While the bulk of technological change is cumulative and continuous, from time to time there do seem to be abrupt discontinuities when, in my terms, one complex is rather rapidly replaced by a rival. These are the events that Constant, following Kuhn, has described as revolutions. He stresses many parallels with Kuhnian revolutions: he argues for example that revolutions are precipitated by the accumulation of some threshold level of functional failures and presumptive anomalies; he emphasizes the abrupt switch in the technological community as the revolution takes place; and finally he points out the problems of maintaining consistent assessment standards across revolutions. Many of the same criticisms that have been levelled against the Kuhnian concept of a scientific revolution can also be laid at the door of its technological analog. Just why should there be some threshold of anomalies and failures that precipitates a revolution when some problems always exist? Is it really the case that one community replaces another as quickly as is claimed? And, whatever the problems, is it not the case that in fact sound judgements are made about the merits of pre- and post-revolutionary technologies? The concept of a technological revolution needs a great deal of further work in order to clarify what seems to be a sound historical intuition about dramatic events that have occurred in the past. All I can do here is offer a couple of comments of a preliminary kind.

One rather frequent way in which technological revolutions start is by what one might call the supplement-redundancy sequence. An interesting example is the replacement of water wheels by rotary steam engines as power sources in the late eighteenth and early nineteenth centuries. Reciprocating steam engines were first manufactured in the eighteenth century and used primarily to pump water from mines. As traditional water-powered industries, dependent on the rotary motion of the water wheel, found themselves short of water, they employed reciprocating engines to pump water from below to above the water wheel, and thus recycle it. From this it was a short step to perceive that *if* a rotative steam engine could be developed an expensive and potentially unnecessary stage in the system could be eliminated. There were already ways of translating reciprocating motion into rotary motion available – in particular, the crank – and consequently this device was adapted for the construction of a rotary steam engine. (The story is *in fact* a little more complicated, since there were several other problems to be solved. Watt, who had the most complete set of solutions, felt he could not use the crank which had been patented by a rival in an engine design, and invented other linkages, including sun and planet gearing. However once his rivals' patent had expired, engines were constructed with the simpler crank).

This example points up another very important feature of technological change in general, and revolutions in particular, namely the importance of the environment in which the technology is to be employed. Often shifts in this environment precipitate the development and adoption of new complexes. Sometimes the environment itself changes, as in the case of the English landscape in the seventeenth and eighteenth centuries when deforestation became a real problem. In other cases the technology may be transferred to a new environment, as when British technology was introduced in America, which was rich in natural resources but short of labor. In both cases the old technology was modified in very significant ways. Indeed it is often the case, as with water power in the nineteenth century, that the most rapid change and improvement in the old technology take place just prior to the revolution. Thus revolutions are not necessarily due to the gradual stagnation of old traditions. Even more important, these episodes show that since revolutions often occur with respect to a specific environment, a technology such as the water mill may be perfectly satisfactory in one environment where water is plentiful, and pitifully inadequate where these conditions do not hold. Technological change often results in the production of technologies and complexes that divide the environment into increasingly specialized niches, each with its appropriate technology. Yet for all that we do want to say that technology has made progress, and that the specialized, complicated technologies are *cognitively* if not socially or environmentally an improvement over earlier, simpler and more widely applicable ones. Just how this cognitive progress is to be spelled out remains an open question.

9. CONCLUSION

This has been a necessarily brief survey of some of the categories that have been developed for understanding scientific change, and their potential applicability to technology. Obviously much more work remains to be done. However it seems to me that there are enough parallels between science and technology to warrant such work. Furthermore if we want to understand the internal cognitive development of technology, as one step toward placing it in its wider context, then this approach has the advantage of building on work that has already been done. In the meantime, if this generates further discussion, it will have served its purpose.

Center for the Study of Science in Society
Virginia Tech

NOTES AND REFERENCES

1. See the introduction to this volume. My approach to technological change echoes that expressed by Nathan Rosenberg and Walter Vincenti who say "The essential nature of technological change has been obscured by an excessive tendency to treat it in vague, general, or purely abstract terms.... Our assumption is that technological change can most fruitfully be examined as a problem-solving activity." *The Britannia Bridge*: *The Generation and Diffusion of Technological Knowledge* (Cambridge, Mass: MIT Press, 1978).
2. See Thomas Kuhn, *The Structure of Scientific Revolutions*, 2nd ed. (Chicago: Chicago University Press, 1970); Imre Lakatos, 'Falsification and the Methodology of Scientific Research Programmes,' in Imre Lakatos and Alan Musgrave, eds. *Criticism and the Growth of Knowledge* (Cambridge: Cambridge University Press, 1970); Larry Laudan, *Progress and its Problems* (Berkeley: University of California Press, 1977); and Thomas Nickles, *Scientific Discovery*, vol. I (Dordrecht, Holland: Reidel, 1981).
3. For just a few examples see footnote one and the following: "It was generally agreed [by participants in a conference on the topic] that inventive activity was a form of problem solving." Richard R. Nelson, in *The Rate and Direction of Inventive Activity* (Princeton: Princeton University Press, 1962), 7; Herbert Simon, *The Sciences of the Artificial* (Cambridge, Mass: MIT Press, 1969), and Thomas Hughes, 'The Electrification of America: The System Builders,' *Technology and Culture* 20 (1979), 124-161.

4. These technologists may be, variously, inventors, mechanicians, entrepreneurs, engineers or members of a number of other occupation groups. I shall refer to them conjointly as technologists in order to avoid confusion, for while the practice of technology has become increasingly professionalized in the last hundred years, there still remain engenderers of technological change who do not really qualify as engineers, while few nowadays are likely to identify themselves as mechanicians. Many, though not all technologists, may also be instrumental in the production, maintenance and diffusion stages of technological development, but those stages need not concern us here.
5. Alvin Weinberg, *Reflections on Big Science* (Cambridge, Mass: MIT Press, 1967), 141.
6. Such standard definitions of technology as "man's efforts to satisfy his material needs by working on physical objects," Charles Susskind, *Understanding Technology* (Baltimore: The Johns Hopkins University Press, 1973), 1, tend to obscure this important point.
7. I take this term from Edward Constant, *The Origins of the Turbojet Revolution* (Baltimore: The Johns Hopkins University Press, 1980), 12.
8. Thomas Hughes, 'Inventors: The Problems They Choose, The Ideas They Have and The Inventions They Make,' in Patrick Kelly and Melvin Kranzberg eds., *Technological Innovation: A Critical Review of Current Knowledge* (San Francisco: San Francisco Press, 1978).
9. Albert O. Hirschman, *The Strategy of Economic Development* (New Haven: Yale University Press, 1958), ch. 6.
10. Nathan Rosenberg, *Perspectives on Technology* (Cambridge: Cambridge University Press, 1976).
11. Constant, *Origins of the Turbojet Revolution*, 13.
12. *Ibid.*, 15.
13. Some scholars would argue that the need for systematicity is an even stronger constraint in technology than in science. See Edward Constant's paper in this volume. This seems to me a questionable claim for which more evidence needs to be adduced.

14. See Thomas Hughes, *Elmer Sperry: Inventor and Engineer* (Baltimore: The Johns Hopkins University Press, 1971), 64-70 for a fascinating account of how at least one practitioner selected his problems.
15. This bears some resemblance to Bertrand Gille's suggestion that the structure of technology at a given time can be thought of in terms of three concepts. First and most elementary is the 'technical structure' – the hammer, the crank, the heat engine. This concept is, I believe defined in terms of function. Second, there is the 'technical ensemble' – a complex of techniques designed for producing a particular product, for example, iron smelting. Each element must be present or the ensemble will be 'blocked'. Finally there is the 'concatenation of technical ensembles' which have considerable interdependence and coherence. Progress during the period of a general system takes place by a system of imbalances. No invention can become an innovation unless the social system permits it. See Bertrand Gille, *Histoire des Techniques* (Paris: Gallimard, 1978).
16. This should not be taken to mean that individual technologies are never regarded as ends in themselves. The Eifel Tower is a wonderful example of 'pure' technology, and a stunning illustration of the fact – so often forgotten – that technology cannot be understood entirely in terms of utilitarian aims.
17. Thomas Hughes, 'The Electrification of America,' The System Builders,' *Technology and Culture* 20 (1979), 124-161.
18. Joseph Schumpeter, *Capitalism, Socialism and Democracy* (New York: Harper, 1942).
19. Lewis Mumford, *Technics and Civilization* (New York: Harcourt, Brace & World, 1934) and Gille, *Histoire des Techniques*.
20. Herbert Simon, *The Sciences of the Artificial* (Cambridge, Mass: MIT Press).
21. Merritt Roe Smith, *Harper's Ferry Armory and the New Technology: The Challenge of Change* (Ithaca: Cornell University Press, 1977) and Anthony Wallace, *The Social Context of Innovation* (Princeton: Princeton University Press, 1982).
22. Nathan Rosenberg, 'Technological Change in the Machine Tool Industry,' in *Perspectives on Technology*.
23. Constant, *Origins of the Turbojet Revolution*, 15.

Derek J. deSolla Price

NOTES TOWARDS A PHILOSOPHY OF THE SCIENCE/TECHNOLOGY INTERACTION

For something usually taken for granted as a basis for practical politics and economics, the lack of critical scholarship on the interaction of science and technology is downright scandalous. All that seems clear is the inadequacy of the naive idea that somehow or other science can be 'applied' to make technology. In spite of a morass of case studies promoted as a legitimation of funding practices or to provide a social license for the support of basic scientists, no such concept as 'application' is of use to serious historians of science or of technology. Similarly, although there are voluminous statistics which disaggregate such entities as funding and manpower into categories of basic science, applied science, and development, there is no evidence that this division produces results of any theoretical value. On the contrary they appear to be nothing but a trivial artifact of the definitions used, rather than any illumination of the chain of action in which they are supposed to be linked.

I suggest that this unhappy position has come about partly through the regrettable though expedient separation of history of science and history of technology, each with its own approach to what constitutes historical explanation. Partly also it has evolved from what now seems to me a misguided main line within philosophy of science in which it is supposed that the function of scientific experiment is the testing of hypotheses and theories. This supposition originated in an atypical period in the early nineteenth century when experiments in electricity and magnetism were common, and it continues to flourish because those who have their experience of science from books rather than from the laboratory bench naturally regard the history of science as if it were an intellectual enterprise, pure and simple. The situation has been exacerbated by the work of Kuhn, who suggests that discontinuities in intellectual history can be regarded as mysterious changes of paradigm, thus preserving the myth that, discontinuous or not, the history of science consists of a flow of cerebral events.

R. Laudan (ed.), The Nature of Technological Knowledge, 105–114.

The burden of this paper is to redress the balance by viewing the history of science as only partly a flow of intellectual steps.[1] The other part is the craft of experimental science which is part of the history of technology. Each radical innovation in this craft tradition gives rise, not to the testing of new hypotheses and theories, but rather to the provision of new information which affects what scientific theories must explain. This process, which I describe as 'artificial revelation', is at the root of many paradigm shifts, perhaps not all, but most. In these cases the paradigm shift comes about because of a change in the technology of science which may be rather trivial and is almost always an intruder from some vastly different current in history of technology.

This is more important than a mere change of current doctrine in philosophy of science. The phenomenon described identifies an automatic link between the histories of science and technology. This link, which would seem to have occurred commonly, may well have been missed because scholars have been looking for a link that proceeded from science to technology rather than in the reverse direction. We should have been warned by the well-known historians' epithet that thermodynamics owed much more to the steam engine than the steam engine ever owed to thermodynamics. Usually when things from the world of artifacts are studied, this is called 'applied science'. The term is a very dangerous misnomer, for many people seem to have assumed unthinkingly that the term implies some sort of application of 'basic science'. It would be possible to stretch a point and say that applied science uses the same techniques and methods, and perhaps also the same logic as basic science. I feel that a much clearer usage is to say that if we research the natural world we get basic science, and if we research the manmade world we get applied science. This sort of connection is, however, much less important historiographically than the other connection to which I have drawn attention, in which a special branch of technology, the technology of scientific instruments, is crucial in making great changes and even paradigm shifts in basic science and maybe also in applied science.

The archetype of this phenomenon is the earliest pathbreaking instance, namely Galileo's use of the telescope. From the viewpoint of history of technology we can see that the telescope became possible when the crucial component – strong diminishing lenses – became commonly available during the late sixteenth century. Lensmaking had been an honorable and common craft for producing eyeglasses in the late middle ages, but lenses for myopic people were a late development, and very strong diminishing glasses had little commercial utility. In the Renaissance, with its preoccupation with artists' perspective and other illusions, such perspective glasses attracted interest since they showed the world in miniature like a microcosm. Once the lensgrinding lathe had been developed for the mass production of eyeglasses, deep-grinding a lens blank became easy and these special highly concave glasses could be simply produced. After they became available skilled workpeople, with the new lenses at hand, could have produced by trial and error the only two-lens combination that produces an interesting result even when the lenses are not adjusted to give correct focusing. There are other two-lens combinations that yield an image – the Keplerian telescope and the compound microscope – but both these give blurred images if not carefully focused. Just why a pair of Flemish lensworkers should have produced the telescope device cannot be determined. It could have happened (and probably did without notable issue) in several other major glass-working centers.[2]

At this point let us note that the story as told constitutes a piece of explanation in the history of technology. Customarily we historians of technology are preoccupied with the preliminary gargantuan effort of translating artifacts into writing, and we become weary before the next stage of welding the writing into historical coherence and explanation. I would also make the point that when a technological discovery is made it has economic value and its development therefore follows a different pattern from scientific discovery which, theoretically speaking, is a free good, since scientific discoveries are automatically disseminated in the process of being released to the scientist's peer group for their validation and evaluation. As technologists, the lensmakers tried to sell their invention as a military device to the richest Pentagon of their age. They were wrong, as inventors usually are, about the utility of their device. It was centuries before the telescope was of use in warfare – and then only for signalling rather than for spying.

When Galileo duplicated the device he had no idea that it was going to be of more interest to him than simply as a way of obtaining a commission from his Medici patrons. The Galilean telescope has such a tiny field of vision – like two keyholes in tandem a yard apart – that seeing anything through it is difficult. Most likely Galileo saw only the Moon on the first night, and was perhaps not much more than amused to see the illusion that the Moon looked as if it had mountains on it instead of the face of a man in the Moon. The *geistesblitz* struck only when he looked again a few days later and saw that the shadows of the illusory mountains now looked completely different from the way they had earlier. Galileo knew enough astronomy to realize that a back-of-an-envelope calculation would enable him to compute the height of those mountains. He carried out the computation and found they were about the same size as mountains on the earth – this gave him what is sometimes called a 'click'; everything fell into place. He knew the mountains were real and no illusion. He knew that for the first time in history he had seen something that could not be attested from common experience of the senses.

The magnitude of this discovery cannot be overemphasized. That the Moon had mountains was an important discovery, but faded into relative triviality when compared with the nature of the experience itself. Galileo realized that he had manufactured for himself a revelatory knowledge of the universe that made his poor brain mightier than Plato and Aristotle and all the Church Fathers put together. This principle of artificial revelation was what was to worry the Church into behaving beyond the bounds of toleration and fair play towards a devout Catholic. It was not just that they could not take such an intellectual threat in the age of political and economic danger from the Reformation. It was certainly nothing so trite as a clash with authority over the specific points of the Copernican position.

To understand the Church position, think of our own rejection of the proposition that ingesting such drugs as LSD gives one access to new data about the nature of the universe. Most of us regard such evidence as the artifact of the mind-bending effect of the psycho-pharmacological agent. Then again think of the distaste many people feel for the idea that an electronic computer might reach the stage where it could contain more information than any human and be taught to reason with it so much better than us that it would not only beat us all at chess, but be able to think creatively beyond the power of humankind. We tend to find it abhorrent that a mere machine should be able to supplant us in such a fashion. Yet, after all, this scenario is but one stage removed from the experience of Galileo. His little tube with lenses clothed the naked eye, allowing it to exceed all previous human experience. Clothing the naked brain is a step still to be taken, if not already being taken.

Moving now from the technological change to the scientific, let us consider the nature of Galileo's actual observations and their effect on theory. Clearly, since he did not know what to expect, he could not have begun his observations with the intent of testing Copernican theory. When Galileo unexpectedly saw such phenomena as the phases of Venus, the satellites of Jupiter, and the enormously great multitude of stars beyond those visible in the naked eye universe, he immediately embarked on a great publicity campaign in the vernacular to disseminate his findings.

These unexpected sights, of unquestioned reality, made it obvious, almost without argument and certainly without mathematics, that the universe was Copernican. The Copernican revolution was not completed by Copernicus himself, but now nearly two generations later, making at one blow Galileo's reputation as well as that of Copernicus. As is well known, until then there had been only a very gradual acceptance of the Copernican system since Copernicus could not clinch his case and no astronomer had reason to accept the new system. New calculations based on the Copernican system gave the same mathematical results as calculations in the old Ptolemaic system. Indeed, Copernicus' new system was rather more complicated than the Ptolemean, and his calculations not necessarily preferable. Kepler did, indeed, clinch the matter before Galileo's telescope, but his work involved such deep mathematics that few if any of his peers and contemporaries ever followed it successfully – certainly not Galileo, whose control of such mathematics was far too poor for any real understanding.

Returning to Galileo, there is little doubt that the resounding success of *The Heavenly Messenger* and of the telescope – everybody wanted one, to 'turn on' and to share in the new way of doing philosophy by artificial revelation rather than by brainpower – made his reputation. Before the telescope, Galileo was not a particularly distinguished natural philosopher. After it, he was a giant. He had 'looked on Nature bare', and learned from the experience that it could be done again and again. His insight was that the new aid to the senses was generalizable to other experiences; this changed thinking to the New Philosophy of experimental science. This was not the *testing* of theories, but the trying out of new techniques to see what they would give, hoping for the unexpected.

My prototypical example focuses on the same revolution that Kuhn takes as a prime candidate for a paradigm shift. Kuhn fixes on a leap of the imagination by a great cognitive master. But I see instead that leap as faltering and almost failing, and the huge and revolutionary change taking place when the technology of the telescope clothed the previously naked eye. While I may agree with Kuhn that subscription to paradigms provides the sociological glue that binds together parties of scientists searching similar territory of science in similar ways, I disagree with him about what causes change of paradigm. Kuhn sees paradigm shifts as being produced by great leaps of the human intellect. Some may have occured. I see the most common cause for shifts in paradigms as being changes in technology, often quite slight and innocent in appearance, but in fact providing new sorts of window for scientists to look from.

I claim a considerable advantage over Kuhn's account of paradigmatic revolutions, and the philosopher of science's interpretation of experiment as a handmaiden for testing theories. I believe that the history of the craft of experimental science is the missing link between the history of technology and that of science. The history of the craft, although none too extensive in its literature, and usually uncommonly antiquarian rather than historical, is nonetheless honorable. Its importance is seminal since it is now recognized that the ancient craft of the clockmaker runs side by side with that of scientific instrument making, and that those twin crafts contained much of the repertoire of fine metal work which made possible the Industrial Revolution. The detailed history of scientific instruments and the later development of laboratory apparatus could probably be linked to general developmental patterns in other technologies, using the same types of explanation that I have given for the origin of the telescope. Local conditions and the economic rule of the marketplace are of the essence. The history of the crafts of experimental science is externalist, rather than the internal pattern of logic with an occasional leap that appeals to intellectual historians of science.[3]

The history of the experimental crafts led to unforeseen observations that changed the destiny of science quite abruptly by their revelations. The most consequential example is the discovery of voltaic electricity, which came about through the researches of an anatomist, Galvani, into the topic of the very nature of life. Galvani believed that the vital fluid must be chemically allied to the electric fluid produced by static electrical generators, because a shock from a generator caused muscles to contract quite violently. He used a muscle isolated from a frog leg, and noticed it could be made to twitch without being connected to the machine; apparently the contact between the hook in the leg and the rail of the bench somehow caused the twitch. The physicist Volta went a step further and showed that the frog leg was not needed at all. The contact between the pair of metals was alone enough to produce a new effect, a new phenomenon. Developing the Voltaic pile of several junctions between discs of metal separated by a pad moistened with weak acid, Volta made what was in effect an electric battery. Once demonstrated it swept the ranks of amateurs of science like wildfire.

Current electricity was not immediately recognized as a source of energy in the modern sense, rather it was thought to be a new sort of imponderable fluid, a chemical juice. As such it was the first new source of chemical change since fire and water, and within a few years the new fluid had produced many new chemical elements by electrolysis. The discovery of highly reactive substances like sodium and potassium allowed the expansion of the table of elements from items known since prehistory to nearly modern form. 'Wet chemistry', inorganic analysis and synthesis, was developed within a single generation. Organic chemistry followed rapidly, as well as Liebig's revolutionary institutionalization of academic chemistry. This had a showcase of products – fertilizers, dyestuffs, pharmaceuticals and explosives – that formed the first group of technologies based on science. Better living through chemistry came out of the attempt to find the secrets of life in the back legs of a frog.

Somewhat later electrical science and electromagnetism were developed in their own right by Faraday, Oersted, Ampere, Maxwell, Edison and others. The early aberrant period of this development, which involved experimentation with electrical and magnetic measuring instruments, gave rise to the myth, prevalent from Whewell through Popper, that experimental science is the activity of testing theories.

Later, the story led rapidly to better living through electricity. The birth of the twin new sciences of chemistry and electricity, both producing immediate technological benefit, is the keystone of the later industrial revolution and the birth of high technology. Industrial benefits did not arise indirectly from an application of chemical and electrical theories, rather they came directly from the same techniques that had produced the theoretical change.

In general, technological change in the craft of experimental science tends to occur rather capriciously, often as the accidental outcome of trying to do something else and noticing a peculiar effect or phenomenon or a strange substance. This may occur in routine scaling up of a technology (the glass lathe), or it may happen when a routine trial is run (as when Rutherford noticed artificial radioactivity when he tested gases for transmission of alpha particles). The new technique may be treated as a trivial toy (like silly putty or Maxwell's stained glass windows visible in color through a polariscope). The effect may be intriguing and pretty (Cherenkov radiation), or immediately recognized as a major puzzle (radio-astronomy), or as a major industrial breakthrough (the transistor). The new thing may be an instrument (the first cyclotrons), or a phenomenon (radio waves). It may be a substance, an effect, a phenomenon, a methodology, or a technique. Even the techniques of calculus and infinite series which were the keys that made possible the great synthesis of Newton can be placed in this category. I suggest we use the term 'instrumentality' to describe these craft innovations of the laboratory.

Such an instrumentality, once discovered or invented, has immediate potential for the generation of new scientific knowledge which may be described as adventitious. The knowledge is not being especially sought but appears gratuitously. With luck it may not fit available theories, causing tension in the theoretical system, and requiring it to bend or even to break. At the same time a new instrumentality may cause instead, or as well, a new technology, a new potential application to the market. Such instrumentalities constitute a corpus of 'almost technologies' which lurk in the world's laboratories and are available to the ingenious inventor and hence to the innovative process.

Significantly the most highly cited papers are 'method' papers, that effectively have the status of such instruments even though they are software rather than hardware. Further the process I have described is not confined to the exact sciences. Darwin's access to the Beagle was an instrumentality. In the social sciences such devices as the national census and the public opinion poll are instrumentalities.

Finally, very simple and almost trivial instrumentalities such as *in vitro* culture of microorganisms may have almost explosive consequences for both science and technology. More often than not the ingenious people with brains in their fingertips who first develop these powerful instrumentalities are unsung, frequently anonymous technicians.

In a previous paper[4] I have suggested that science and technology move in linked but independent ways, related like a pair of dancers. Now I argue that what keeps them linked is that both dance to the music of instrumentalities. Normal science begets more normal science. Normal technology begets more normal technology. But an adventitiously new instrumentality can make for a change in the paradigm within science, and an invention leading to new innovation within technology.

History of Science
Yale University

NOTES AND REFERENCES

1. This is a short version of a more documented and elaborate article, 'The Science/Technology Relationship, the Craft of Experimental Science, and Policy for the Improvement of High Technology Innovation,' in *The Role of Basic Research in Science and Technology: Case Studies in Energy Research and Development* (Division of Policy Research and Analysis, National Science Foundation, 1983).
2. Albert van Helden, 'The Invention of the Telescope,' *Transactions of the American Philosophical Society* 67 (1977), 67 pp.
3. Derek J. deSolla Price, 'Philosophical Mechanism and Mechanical Philosophy: Some Notes Toward a

Philosophy of Scientific Instruments,' *Annali Dell'Istituto e Museo di Storia Della Scienza di Firenze*, V, (1980), 75-85; see also G. L'E. Turner, 'The London Trade in Scientific Instrument-Making in the 18th Century,' *Vistas in Astronomy, 20* (1976), 173-182 and 'Apparatus of Science in the Eighteenth Century,' *Revista da Universidade de Coimbra* 26 (1977).

4. Derek J. deSolla Price, 'Is Technology Historically Independent of Science? A Study in Statistical Historiography,' *Technology and Culture* 6, (1965), 553-568.

Peter Weingart

THE STRUCTURE OF TECHNOLOGICAL CHANGE: REFLECTIONS ON A SOCIOLOGICAL ANALYSIS OF TECHNOLOGY*

1. INTRODUCTION

When science became an issue of public concern in the sixties, research activities in many fields and notably in sociology which were directed at a better understanding of its dynamics and relations to the social environment flourished and have resulted in the emergence of a strong sociology and social studies of science. Now that technology has become a public issue, sociologists have hardly anything to contribute to the understanding of its dynamics. Technique, technology, or more generally artifacts, have had no systematic place in sociological theory since modern theory of action superseded Marx and Durkheim.[1] An effort to develop a sociology of technology, therefore, is timely, the assumption being that the many attempts at explaining the dynamics of technical development which have been undertaken by economists, philosophers and historians of technology, may be improved by adding a sociological perspective.[2]

If a sociology of technology seems a feasible venture, the problem before us is very much like that of developing a sociology of science, both science and technology being systems of knowledge evolving in structures of social action. The road recently taken by the sociology of science into a kind of neo-internalism may have its merits in elucidating the fine-structure of knowledge production in science.[3] It can hardly be the base from which to embark on the analysis of the production of technical knowledge which is far less self-contained and much more subject to the orienting principles institutionalized in those systems in which technical knowledge is realized as artifacts. Thus, I want to claim that it is necessary to take a step back and approach the problem by applying the basic elements of a theory of action to the production of technical knowledge. Drawing on an earlier attempt to explain scientific change in

R. Laudan (ed.), The Nature of Technological Knowledge, 115–142.

terms of a sociological theory of action this approach which was heavily influenced by Kuhn's 'theory' of scientific development[4] (if one may call it that) can be used for heuristic purposes in the analysis of technological development.

At this stage this approach is exploratory and speculative, and certainly eclectic *vis à vis* the history of technology.

2. ORIENTATION COMPLEXES OF THE PRODUCTION OF TECHNICAL KNOWLEDGE

Taking science as a heuristic model for the analysis of technology it is fairly commonplace to say that the production of scientific knowledge is a structured social activity. The structures are organized around various kinds of orientation complexes with different degrees of institutionalization and ranges of validity. Such orientation complexes may be very general codes of professional ethics, more restricted convictions about the delineations of disciplinary subject matters, still more narrow definitions of areas of research, theories, models etc. or very limited problems guiding small groups in their venture into unknown realms of knowledge. These orientation complexes are not fixed and abstract but substantive ideas and subject to revision. Also, actors are oriented by a variety of them, they overlap or are combined in various ways, and their combination may change as the action evolves. Thus, while it is certainly a simplification to speak of *a* 'scientific community' the relative coherence and structure observable in the communication and research process of science remains to be explained. The crucial questions, then, are: who are the actors who produce technical knowledge, in what types of groups or 'communities' are they organized or what are the orientation complexes that constitute them, how are they institutionalized and how, in this framework, can one account for technological change?

With this approach some assumptions have been made which should not be too hard to accept:

- Technology is a cognitive system. Even technical artifacts are solidifed cognitive systems;
- Technology is thus also a social product and part of a social system. From a simple tool to a complex system of machines and their infrastructure it is constituted by and structures action;

- Technology is not identical with science or even applied science but there are numerous connections and the borderline is blurred. Thus, Skolimowski's delineation: "In science we *investigate* the reality that is given; in technology we *create* a reality according to our designs"[5] is wrong. To define a demarcation criterion is itself a problem which will not be dealt with in this context, however.

What, then, are the 'orientation complexes' of technical research (invention) and development?

That science is characterized by the search for truth is an oversimplification. However, the obvious difference between (academic) science and technology which is apparent both on the level of their institutionalization and on the meta-level of their historical and philosophical reflection can be attributed to the different character of their orientation complexes. While academic science tends to be identified with the search for truth, technology is characterized by the pursuit of ends, no matter where they originate.[6] On the surface technology provides a much more diffuse and amorphous appearance. The reason is that the orientation complexes of technology are multidimensional: in addition to being technical and scientific they may be economic, political and cultural. This is a reconceptualization of a well-known fact.

Johann Beckmann, the Göttingen economist, stated in his *Entwurf der allgemeinen Technologie* (*Outline of the General Technology*) that the general technology aims to enumerate "all the different objectives, which the craftsmen and artists have with their different projects and a catalog of *all means*, with which they know how to meet each of them."[7] Another example is the characterization of the stages of technical development by Alois Riedler, the eminent German machine designer. He differentiated between the development of a machine that *works* (is *gangbar*), a second stage of development of a machine that is *useful* (*brauchbar*), and a third stage, when the machine has to be developed to be *marketable* (*marktfähig*).[8] Beckmann acknowledges the diversity of objectives and the means to meet them all of which may be catalogued to make up the 'general technology'. Riedler's stages of development actually imply that machine development is oriented to scientific, technical and economic constraints and objectives. A few illustrations of the orientation complexes shall clarify their meaning.

Starting with the non-technical orientation complexes, the most important one for technology at least since the nineteenth century is that set which is institutionalized in the economy. The seemingly overriding impact of economic incentives on technological development must be specified, however. The market is the clearing mechanism for the selection (acceptance or rejection) of technologies but does not determine directly the development of new ones. Rosenberg notes that "economic incentives to reduce cost always exist in business operations, and precisely because such incentives are so diffuse and general they do not explain very much in terms of the particular sequence and timing of innovative activity."[9] Cost considerations do enter the decision-making process of construction as a *general parameter* and determine the design process in a general way.[10] The priority of the criterion of 'economy' is documented in the basic principle of construction which is not "as good as possible" but "as good as necessary".[11] Attempts at enumerating parameters of construction (which in action-theoretical terminology are orientation complexes) have led to catalogues of varying length. It can be shown, though, that the choice of particular values among such criteria as durability, speed, efficiency, work load etc, may very often be traced back to the cost factor and is specific to the technique in question. That shows that economic orientations do not only have a selective function between technical possibilities but enter the construction process from the very beginning. But in order to be a determining orientation it has to be operative in a specific technological and industrial or some other context. Also, it will still have to be translated into constructive solutions which are subject to technical-physical principles.[12]

A particularly fitting example is, of course, Edison's approach to the development of an incandescent lighting system. Trying to replace gas lighting he not only computed the costs of that system to the customers which had to be met in order to make the new system competitive. He also had to take the costs of copper as a parameter which determined the choice of the voltage of the electrical system and, as a consequence the search for a sufficiently durable material for the incandescent filament. In other words, Edison was aware of the selective power of the market which oriented him to the extent that he had to design his lighting system in a way that consumers would not discard it on the basis of price considerations. The invention of electric lighting can thus be considered as economically determined.

A second set of orientation complexes emanate from the political system. Political orientation complexes can be selective or determing as well, and they may be combined with economic ones, or they may exclude them. Examples of selective criteria are safety standards of all kinds although they may determine technical development as well and they may also change the economic parameters involved. Air pollution standards or standards of noise emission (in practice mostly geared to 'the state of the art') may induce specific development processes but they are selective with respect to different technical solutions. To put a man on the moon within one decade is a political orientation which determines the direction of an entire system of technical development. To 'humanize' working conditions is a rather vague orientation which needs operationalization but may end up at least in partial conflict with economic criteria. Political orientation of technical development has a long tradition in military engineering and here there is neither a market (except in peacetime when weapons are built for export, too) nor do cost considerations enter prominently into the construction process.[13] A new set of inherently political orientation complexes emerges in the form of 'technology assessment'. Social acceptability becomes a general criterion of technical development which is concretized in the attempt to anticipate public response to particular technologies.

A third set of orientation complexes may be called cultural. This category is added to draw attention to the vast realm of technological developments which are oriented by aesthetic, religious and ideological criteria. The installation of waterworks in the royal gardens at Versailles under Louis XIV contained advanced technology designed for pleasure. Before utilitarianism took hold throughout the sixteenth and at the beginning of the seventeenth century, artist-engineers and architects designed (and sometimes realized) models of pneumatic clockworks, pumps, fountains and other 'curious machines' which reflected their feudal employers' fascination with mechanics and its merger with decorative, aesthetic and hedonic ends.[14] Cyril Smith has drawn attention to the fact that techniques of dealing with materials and the search for their properties were inspired, above all, by artistic, i.e. aesthetic motives. Likewise he points out that "the relation between design, structural engineering, and knowledge of materials in architecture is a well-known example of the inseparability of aesthetic and technological factors ... that it has usually been nonutilitarian structures such as temples and monuments that have stretched the limits of existing techniques and led to the development of new ones."[15]

Another example is the influence of ideological orientations on architecture, both utilitarian and non-utilitarian structures. Albert Speer himself points to the French revolutionary architects such as Boullées and for a short time fascist ideological principles won over the established architecture, in that case not stretching the limits of existing techniques but reaching back to constructive elements of the past.[16]

Finally, technology as a cognitive system provides an important orienting frame of reference. (For the sake of brevity I include science as part of that system but want to avoid the discussion of the relation between science and technology and its change over time. Insofar as scientific knowledge is the basis of technology it may here be subsumed under technical knowledge.) It is a truism that existing technical knowledge is the orienting base from which new technical knowledge is developed. To some extent it structures the perception of technical problems and their solutions.

The establishment of schools of engineering, with curricula for the various branches of engineering, suggests the formalization of technical knowledge so that it can be taught in abstraction from concrete ends. This development must be traced into the early nineteenth century when engineering began to emulate the ideal of science. Attempts at establishing a general technology (which were cited above) or the work of Reuleaux who tried to introduce the deductive method into machine engineering in order to achieve a planned 'scientific innovation', or finally Redtenbacher's program to base "*das ganze Maschinenfach auf sichere Regeln*" are all directed to a 'scientification' of technology, to the emancipation from science and the establishment of engineering as an independent knowledge system.

The *École Polytechnique* gave priority to theoretical teaching in geometry, algebra and calculus in order to make its graduates capable of coping with any task.[17] It is interesting that the actual institutional emancipation of technology from science was achieved by abandoning the attempts at scientification. It was Riedler who pointed to the uniqueness of technique and the "diversity of practical conditions" which required unique methods, namely experiments on machines at real scale and under realistic conditions.[18]

These examples seem to point to an 'autonomy' of the development of technical knowledge and consequently to technical orientation complexes independent from non-technical ones. Most pertinent for the illustration of the relative impact of such purely technical (and scientific) orientations *vis à vis* the non-technical ones are cases of parallel development of a technology which take place under the influence of the different orientations. One such case is the evolution of the turbine in France and the United States. While in France the mathematical tradition of the *École Polytechnique* prevailed which led to highly sophisticated designs and had Fourneyron develop his theory of turbine construction, in the United States the principal orientation was economic, exemplified in the practical craftsmanship of millwrights. Their turbines were less sophisticated but cheaper and adapted to the contexts of their use.[19]

With the appearance of science-based technologies (or technology induced by science) it is obvious that technical and scientific orientation complexes may have an independent role. A recent study of the origins of the turbojet engine seems to show just such a development, where scientific theory suggested the possibility of higher aircraft speeds and higher compressor and turbine component efficiencies.[20]

This case points to a pattern of technological development (one among others) in which the basic concept of a technology emerges in the realm of scientific/technical orientations and only in the course of their realization as concrete artifacts do other orientations enter into the process: in the case of the turbojet engine, military objectives and specifications and later on economic ones. In other words, even in cases where scientific/technical orientations have an initially independent impact they are mediated by other orientation complexes in the design process. It is then that 'simplifying assumptions' and compromises are applied as is exemplified in the engineering methodologies of parameter variation or dimensional analysis.[21]

The autonomy and independence of scientific and technical orientation complexes in the development of technologies *vis à vis* the non-technical ones raises the well-known issue of a demand-pull versus a supply-push model of technological innovation, as is termed in the literature. While one need not deny that technical knowledge becomes formalized and systematic – as is documented in the 'engineering sciences' – it "is not a body of systematic knowledge comparable to that of science"[22] in the sense that it is always developed in *conjunction* with other non-technical orientation complexes.

If the concrete technology (or technical system) is taken as the frame of reference, as it has to be, it comprises a multitude of orientations or elements of knowledge from a diversity of 'bodies' or sources: in turbomachinery for example there are bits of elegant science, mechanics going back to Euler and Bernoulli, fluid mechanics, simplified schema like velocity triangles, techniques of analysis such as dimensional analysis which involve a wide range of choice in selecting the dimensionless parameters of numbers to be studied, and purely empirical data and formula.[23] Thus, in a sociological sense technical knowledge is *amorphous* and the dichotomy of demand-pulls versus supply-push models is misleading. As will be shown below technological change must be interpreted in terms of the impact of varying compositions of orientation complexes on the production of technical knowledge.

3. THE INSTITUTIONALIZATION OF TECHNICAL ORIENTATION COMPLEXES

Having said that technology is both a cognitive and a social system this leaves open who the actors are and how extensively the orientation complexes organize action. In the analysis of science this seems to be a relatively simple problem if, following Kuhn, paradigms are said to constitute community structures such as specialists' groups, the connecting mechanism being socialization both in formal education and practice. This direct identification of cognitive and institutional structures has long since then been abandoned up to a point where some authors want to dissolve such a connection altogether.[24] I want to make a case for retaining the analytical approach in the study of technology as a particular type of knowledge while at the same time looking for adequate differentiations. This implies identifying the forms of institutionalization of the orientation complexes enumerated above, for in order to become orientations of knowledge production in a more than accidental and passing fashion they have to acquire some persistence and stability.

Economic orientation complexes in their most general form are institutionalized in the market (wherever there are market economies) or in economic plans. If it can be said that they constitute patterns of social action and thus institutional structures it is the planning staffs and development laboratories of firms which, as one strategy among others, try to innovate in order to gain an advantage over their competitors. While such general criteria as a cost-benefit ratio guide all firms and are therefore unspecific, they are being translated into concrete programs of production. Not the market as such, but that for computers or automobiles, constitutes an institutionalized network of communication and it is the group of engineers in that particular branch of industry as well as the salesmen, marketing staffs, managers and others, all of whom work for, with and around the particular product, who are connected by this network.

It is frequently argued that the corporation is the relevant institution for the initiation and channelling of technological development and thus the principal agent giving institutional weight to economic orientation complexes. This has to be qualified as the diversified large-scale concerns especially are not committed to developing certain technologies but to making profits with the capital of their shareholders. They are the institutional locus of general economic orientation complexes which, above all, have selective functions, identifying new technological opportunities, opening up markets for them, providing the capital base etc. On the other hand, the traditional type of corporation, organized around one particular technology (e.g. the railroad company, or today the genetic engineering firms) which emerged with and often disappears with that technology, can be said to be the institutional locus of economic orientation complexes whose impact is specific with respect to that technology. It may be added that the forms of institutionalization are subject to historical change, of course, of which the development from the small manufacturer to the multinational concern is the best exemplification. Obviously, each of these, and the intermediate forms, imply different strategies of innovation with different effects for technological development.

The institutionalization of political orientation complexes has a long tradition. A particularly pertinent example is the institutionalization of civil and military engineering primarily in the late eighteenth and early nineteenth centuries, when special schools were established (the various *Écoles* in France, the *Bauakademie*, *Bergakademie* and others in Germany) which virtually guaranteed state employment for their graduates and whose curricula were closely geared to the needs of the state.[25]

Another form of institutionalized political orientation complexes is the ever growing body of regulations governing standardization and testing of technologies in order to assure safety, quality, performance, protection of the environment and other objectives. These standards are defined by political administrations whose need for information has given rise to scientific and technical research installations, institutionalized as a branch of government research and development, e.g. the Bureau of Standards in the United States, or the *Bundesamt für Materialprüfung* and the *Physikalisch-Technische Bundesanstalt* in Germany and similar institutions elsewhere.

The communication networks which are constituted by these institutions and which translate the political orientation complexes into technological development typically comprise the engineering profession and various branches of public administration.

I will not elaborate on the institutionalization of cultural orientation complexes, except to point out that economic (and to a lesser extent political) orientations have gained supremacy. This is documented by the fact that function and style are amalgamated. If one looks for examples for institutionalized cultural orientation complexes operative in technological development it is perhaps primarily the prizes and evaluating mechanisms for industrial and architectural design which come to mind. Of course, one may also consider institutions such as the sports industries with their vast number of differentiations as the organizational framework for hedonistic orientations which – even though being clearly ruled by economic orientations as well – have an identity of their own.

Finally, I want to turn to the institutionalization of the technical orientation complexes proper. Much of the history of technology focuses on the inventor as an individual. This is justified on two counts. Historically the independent inventor, often an amateur, played an important role until as late as the turn of the century.

Analytically, "invention can only come at the hand of some sort of inventors."[26] In terms of sociological analysis, however, the inventor represents a historically transient figure before invention became institutionalized as an organized activity. To avoid a misunderstanding: this is not to deny that even today we find the individual inventor applying for patents, but he is no longer the prevalent 'type' representing inventive activity. Rather, it is being suggested, we view his work "as a complement and supplement to the work of the large corporate laboratory."[27] It is for those reasons that the focus on individuals and their motives as well as the biographical circumstances of their work – although by no means without value – is of very limited importance for the analysis of development patterns and dynamics. For that one will have to look for social formations.

The most important form of institutionalization of technical orientation complexes is the engineering profession and its organizations. Hortleder has shown that for the one and a quarter centuries of its existence, the *Leitmotiv* of the VDI (the Association of German Engineers) has remained virtually unchanged: "The association aims at an intimate cooperation of the intellectual powers of German technology for the mutual stimulation and development *in the interest of the entire German industry*."[28] It is a peculiar feature of the VDI (and other engineering societies as well) that it was founded for the promotion of technology and not for the promotion of the professional interest of the engineers. The connection of that goal with the interests of industry seemed 'natural', and in the founding phase the association also readily identified technical projects with political interests such as the unification of the German empire. The membership structure of the VDI reflects this amalgamation of orientations as well. From the beginning industrialists and higher level managers from industry have been members of the VDI. Although, Hortleder concludes, there is a trend toward giving priority to professional interests, the association's proximity to industry is also unchanged.[29]

On the level of its program or constitution a professional organization like the VDI represents the stable institutionalization of very general orientation complexes only. In the case of the VDI close institutional bonds with industry exist on lower levels of the organization of knowledge production and communication as well. Social structures organized around specific programs in technology, say nuclear reactor or biochemical engineers, comprise formal links to industrial and government administration.[30] Technical education as the chief socialization agent for engineers also reflects the orientation to industrial or government practice, and the long debate over the merits of introducing elements of the humanities into the institutes of higher technical education mirrors the social ambivalence of the engineers. The same links to respective fields of practice which I postulated on the level of the orientation are reflected, not surprisingly, on the level of their institutional structures. The internal organization of the engineering profession, therefore, underscores Layton's observation: "unlike science, technology cannot exist for its own sake."[31]

Looking for the institutional representation of technical orientation complexes is another way of trying to identify the correlate of 'communities' in science. The professional societies are but one indicator. Constant, in a rare attempt to apply the concept of community to technology has pointed to others, such as "identifiable industry, sector and product groupings" or as a crude measure, the Standard Industrial Classification Codes. Constant also gives an explanation for what I call the sociological amorphousness of the technological communities in his case study of the aeronautical community. As a first reason he mentions the hierarchical structure, meaning that that community at large is composed of manufacturers, civil and military users, governmental and community agencies and of "industry, government, private non-profit, and university-related aeronautical sciences organizations." Below this level there are specialized communities of practitioners, like manufacturers of airframes, power plants, or accessory systems. "Thus, the aeronautical community is composed of a multilevel hierarchy of subcommunities." Secondly, Constant points out, community memberships may overlap on any level within the hierarchy. Among the piston aero-engine manufacturers, some belonged to corporations with commitments in the automobile industry while others were also airframe producers. Consequently, Constant concludes, the definition of a community of practitioners such as that of "piston aero-engine manufacturers involves a certain arbitrariness."[32]

These examples suggest that those institutionalized groupings which may be construed as communities represent a complex structure of orientation complexes. Although there are communication structures which are organized around narrow technical problems, methods, or practices, they are interrelated closely with those structures representing the 'interface requirements', i.e. the orientations of other technical and non-technical areas of practice. "For example, the practice of aero-engine makers was bounded by the power, weight, fuel consumption, and reliability expectations of the airframe community, which in turn were shaped by the values and expectations of military and civilian users, whose own desiderata were partially determined by standards maintained by still other communities – the availability of certain fuels and long runways, for instance."[33]

'Technical communities' therefore resemble very closely those social structures that we see emerging at the interface of government, industry, and science as a consequence of comprehensive 'science and technology programs'. We have called them 'hybrid communities' to differentiate them from the traditional academic communities which become part of them.[34] In fact, 'technical communities' seem to be their historical precursors. This difference in structure between the social organization of science and technology is highly important when it comes to the analysis of change because of the diversity of orientation complexes, their respective social systems and the affiliated power resources involved.

4. PATTERNS AND MECHANISMS OF TECHNOLOGICAL CHANGE

From what I have argued so far it is evident that the structure of technological change cannot be quite the same as change in science. Just because technical knowledge is inseparable from the orientation complexes based in other social subsystems a change in any one of them affects some or all others. Technological change is in this sense inherently systemic in nature. As the subsystems involved are functionally differentiated the mechnisms that bring about change are, too. Politics are organized around the medium of power, the economy is organized around the medium of money and in the case of science it is truth. With this in mind I want to explore some aspects of technological change.

The first question is, what are some of the implications of the heterogeneity of cognitive orientations and the requisite institutional settings of the production of technical knowledge for the patterns of technological change?

The second half of the nineteenth century may be seen as a period when the independent inventor is still the most important actor in the process of technological development, but he begins to systematize and direct his inventive activity, either by setting up his own laboratory or manufacturing plant or by taking his patents to established industries. At the time when Otto had developed his gas engine and Diesel started working with his idea of making combustion in an internal-combustion engine take place at a constant temperature, "hundreds of inventors were trying to find a substitute for steam that would not need such an elaborate and expensive system, they were seeking a working fluid that would make possible more efficient or more economic power plants, particularly in small sizes."[35] Improvement of the "scandalously low" thermal efficiency of the steam engine ("well under 5 percent") seems to have been a widespread cause for otherwise unrelated inventors, and the feasibility of the internal combustion approach had been successfully demonstrated by Otto's silent gas engine.[36] Whether that approach constituted a community of, say, 'internal-combustion engineers' is not clear, although in the engineering sciences various publications appeared like Köhler's *Theorie der Gasmotoren* (1887) or Grashof's *Theorie der Kraftmaschinen* (1890). However, if communities existed as communication networks they had little to do with the realization of the technology which, like the Diesel-engine, "was clearly much too large a project to be developed by a single man in a home workshop."[37] As a result inventions typically began to be developed in organizations (industry, laboratories) with different functional references.

Independent inventors like Otto (an autodidact amateur) and Diesel (an educated engineer), thus, became partners or employees in machine factories. For some like Edison or Sperry the framework is their laboratory, for others like Siemens, Kettering or Linde it is the industrial plant they set up to develop and market their inventions.[38]

The fact that, at the time when the independent inventor still played a prominent role, the institutional framework for invention was still diffuse accounts for particular patterns of inventive activity. Independent professional inventors like Edison or Sperry, who were not bound to specific industries and their respective product markets, worked in a variety of fields and sought to apply technical concepts to very different areas of application. Sperry tried to sell his first gyro to Barnum and Bailey Circus. (Hidden in a wheelbarrow it was to stabilize a clown walking a tightrope.) After failure there, he developed a gyro to stabilize automobiles and only then did he turn to ship-stabilization. One-time inventors like Otto and Diesel, once they got their patents issued, entered industrial firms which provided the necessary capital for development of their original inventions until they became marketable. Yet, the examples demonstrate the transition to more coherent institutional structures of invention. In the large industrial concern, with its own laboratory, the range of applications is more or less fixed and determines the direction of research. Thus, Bell Laboratories decided that work in solid-state physics would be fruitful for the improvement of communications technology, and concrete results such as an amplifier and improvements on rectifiers, thermistors etc. were in focus.[39] A further development in this direction is exemplified by industries that did not operate within the framework of an historically grown technology. "These industries have developed various forms of *systems engineering* that provide overall technico-economic guides to the assemblies of components on which the industry depends."[40] In other words, innovative strategies are now characterized by the development and combination of components for different objectives. This evolution of forms of institutionalized 'areas of application' has a correlate in innovation as a systematized activity.

While these examples focus on private enterprise as the central institutional context of inventive activity, the other very important one is government. Governments are agents of technological developments in a variety of ways. Military and civilian technology is being developed in government arsenals and laboratories or by subcontracts to universities and non-profit research institutes, as well as to industry.

The differential impact of economic and political (governmental) orientations in the promotion of technological change is, of course, common knowledge.[41] For governments the attainment of political objectives has the highest priority, while costs, efficiency and profitability play a secondary role. Thus, in the case of civilian technologies, governments may decide on infrastructural measures such as the supply of nuclear energy in order to achieve partial or total independence from other energy sources (such as oil) – which is a political objective colored by military considerations. Consequently, they may promote the development and deployment of advanced technologies (e.g. the fast breeder or reprocessing plants) at a stage when they are 'workable' but not yet 'marketable'. The government creates the market for the new technologies itself and also determines cost thresholds. In the case of military technology this pattern is even more pronounced as the specifications for particular weapons systems are defined by the military bureaucracies within the framework of their strategic conceptions.

With governments increasingly assuming an active role in science and technology policy it must be subsumed that technological change is driven, to a growing extent, by political objectives, i.e. government demand, rather than by commercial market demand only. (Of course, governments have played the role of innovator long before the appearance of the modern corporation.) Incidentally, one feature characteristic of recent large-scale technological systems has to be explained in this context: they are implemented with little regard for their public acceptance. The alliance of government bureaucracies, engineers and private corporations – the latter acting as quasi-public agencies by being subsidized directly or indirectly – circumvents the market and operates through the medium of political power. Consequently, non-acceptance of such technologies by the public can only find its expression in political resistance, leading to legitimation problems with grave political rather than mere market failures. If innovations are introduced by private corporations operating through the medium of money, i.e. profitability, acceptance is assured as much as possible by market testing prior to implementation, and if that fails the consequence is the loss of capital and possibly the demise of the corporation concerned.

From these few illustrations, to which many more could be added, some generalizations may be tried. The institutional contexts which provide the orientation complexes guiding the production of technical knowledge have their own functional logic. The politics of different branches in the public administration like the health services or the armed forces, or the strategies of industrial branches such as the oil, automobile, agricultural machinery or electronics industries and, of course, the science and technology policies that mediate between or connect them are expressions of such 'logics' and they influence the direction of technical change. The functional logic of diverse institutional contexts constitutes the problem horizons for which technical solutions are to be sought and, by implication, defines what is beyond those horizons. It is this circumstance which accounts for the institutionally-specific, uni-dimensional dynamics of technological development. Shifts in technologies (to avoid the term revolutions) typically emerge outside the existing institutional structures, and lead to the creation of new ones.[42]

Turning now to the mechanisms of change residing in the system of technical knowledge proper, it is helpful to recall that in the analysis of science many models and concepts have been developed to describe and explain the dynamics of scientific development, e.g. Popper's refutation principle, Kuhn's paradigm crisis and revolution, and Lakatos' progressiveness and degeneration of research programs, to mention the most prominent ones. The question is if similar mechanisms can be found to operate in the change of technical knowledge.

One way to approach this is by looking at the ways in which problems are identified in the inventive process. It happens that there are some authoritative theoretical statements on this issue by historians of technology as well as a systematic analytical comparison.[43] Hughes compares three approaches: Usher's which focuses on the act of insight and is influenced by a 'gestalt-psychological' model but is linked to action theory; Gilfillan's which is sociological and focuses on technical and non-technical factors invoking inventions, and his own which focuses on the evolution of technical systems. All three approaches have some elements in common which can be directly translated into my own action-theoretical framework.

Usher attributes the inventive insight to the careful study of an existing technology and the discovery of an 'unsatisfactory pattern'. (In many cases the act of insight may "do no more than set the stage for the achievement of the solution.") Thus, Watt came across the basic idea of the separate condenser after "careful study of the performance of the Newcomen engine," which proved that engine to be "a very ineffective means of utilizing the heat energy produced by the fuel."[44] Note, that the analysis which revealed the unsatisfactory pattern implied a criterion of effectiveness which must be related to "function, efficiency, or some other external factors," as Hughes rightly observes.[45]

Gilfillan, not unlike Usher, attributes inventions to inventions themselves, which "are a major cause of changes in the milieu out of which is compounded the inventional complex ...." In addition, however, he sees inventions caused by such non-technical changes such as growth of "wealth, education, population, industrialism, and commercial organization."[46] Gilfillan, thus, explicitly widens the frame of reference for the analysis of the factors accounting for the emergence of imbalances of existing technologies to changes in the non-technical factors which interact with changes of technological systems.

Hughes' analysis, although being very close to Usher's approach, focuses on technological systems as the orienting frame of reference. The act of invention comprises the identification of the 'weakest point' which appears as a 'reverse salient' in an expanding technological system, the 'deep involvement' in the study of such evolving systems and the development of the invention through adoption "to increasingly more complex environments."[47]

In all these approaches, and supported by the case histories to which they are applied, a striking difference between science and technology with respect to the perception of problems becomes evident. If we follow Kuhn, the dynamics of scientific development can be attributed to 'crises' of orienting paradigms which are brought about by the process of elaboration, experimentation and testing when it turns out that the paradigm entailed an incorrect view about nature. Technologies, on the contrary, are realized with reference to set tools, and they cannot contradict the laws of nature. There can be no such thing as a 'crisis' of a successful technical concept or system, except with reference to non-technical orientations. They structure and guide the perception of problems. One mechanism of change, therefore, the essence of inventive activity, is the improvement of 'unsatisfactory patterns', the rectification of 'imbalances' or the identification of 'weakest points'. Very often the case histories from the age of the heroic inventors disguise the fact that this activity necessarily has a dual orientation: one is to the technological system and the other is to objectives such as efficiency, economy, performance, comfort, etc. which can all be traced back to non-technical orientation complexes.[48]

An indirect proof of that thesis is the fact that new technologies do not necessarily replace old ones merely because they are 'wrong', but depend, for example, on economic considerations. Edison's low voltage direct current distribution system continued to be used even after introduction of the alternating current system. Due to the invention of the rotary converter, it was coupled with the new high voltage system which enabled the utilities to keep their heavy investments in urban direct current until they could be abandoned due to age and amortization.[49] Despite the invention and diffusion of the steam ship which had been substituted for the sailing ship in passenger travel as early as 1850, the sailing ship went through a great number of adaptations and technical improvements and made up about twenty percent of the gross tonnage of the U.S. merchant fleet as late as 1913.[50]

Constant, in applying Kuhn's notion of anomaly, elucidates the operation of cognitive dynamics as a mechanism of change, but this, as I will show, involves non-technical orientations, too. He speaks of "presumptive anomaly" in technology "not when the convehtional system fails in any absolute or objective sense, but when assumptions derived from science indicate either that under some future conditions the conventional system will fail (or function badly) or that a radically different system will do a much better job."[51] Constant gives as an example three scientific assumptions which appeared in the area of aerodynamics during the late 1920's. First, that by proper streamlining aircraft speeds could be increased by a factor of two from the then current two hundred miles per hour, which would make jet or reaction propulsion feasible. Second, the application of aerodynamic theory induced a few specialists to believe that vastly more efficient gas turbines could be built. Finally, with near-sonic speeds becoming potentially possible, the propeller as part of the conventional piston-engine concept would fail to operate.[52] The example is one of a science-based technology if one considers aerodynamics a science. Even if it is considered an engineering science, as Constant does, the important point is that the dynamics which create the presumptive anomaly are in the cognitive system itself. Knowledge is being developed along established parameters. Due to that, possible configurations of the system are predictable, including the impossible ones like retaining the piston engine and propeller and flying at super-sonic speed.

Also, potential technical solutions can be envisaged, like the turbojet engine. But the existing technology had not failed in any way, a lot of development actually followed. Thus, there was no actual crisis but only clearly delineated limits to the development of the conventional technology. That development, however, is motivated by non-technical orientations: increase of speed and efficiency. Thus, leaving aside the questions whether talk of a (presumptive) anomaly is justified and whether it constitutes a 'crisis' in the Kuhnian sense, it is important to note that, of course, the system of technical (and scientific) knowledge has its own momentum due to which a multitude of 'potential technologies' are being projected or limitations to existing ones are being defined. It is then a question how and when non-technical orientations enter this process. In some cases they act as parameters very early on (e.g. in the quest for a fusion reactor), in others they may even be shaped by the emergence of technological potentials for which there is no 'market' yet (e.g. electronic games).

Apart from the dynamics of the production process of technical knowledge constituted by the 'functional logic' of diverse institutional settings another 'mechanism' of technological change is what Gilfillan summarily calls "social forces" such as "growths of some all pervasive characters of our civilization".[53] In another terminology they are 'constraints' that suddenly emerge. The wars that cut off France from Spanish barilla, Germany from Chilean nitrate and Russian oil, and the United States from natural rubber have in each case drastically changed the economies of the respective resources and their use and thus induced technical developments which eventually replaced them.[54] In other words, because inventive activity is oriented to economic and political parameters, each technological system is an embodiment of a historically specific set of such parameters. Consequently any change in their configuration creates inducements for further inventive activity. They are externally produced 'crises', imbalances, unsatisfactory patterns or weak points of existing technologies.

I cut off the analysis at this point. Any attempt to push the classification further runs the risk of becoming too general. However, for the purpose pursued here it suffices to show that the analysis of technological change in an action-theoretical framework must extend from the micro-level of the individual act of problem definition and invention to the macro-level of social systems in which the configurations of different orientation complexes change both in dependence on and independently from technological innovation.

5. CONCLUSIONS

First of all, even though I have presented what amounts to little more than a classification scheme to which many refinements can be added, it is obvious that a shift in the analysis which conceives technology as knowledge and focuses on invention as an activity of knowledge production can unify the hitherto diverse studies of the history of technical artifacts, biographical studies of inventors, studies of the behavior of the maximizing firm etc.[55] While all of these approaches have their merits within their respective disciplines and focus on different aspects of the process of invention, their flaws appear when they are generalized. This is particularly annoying when generalizations lead to mystifications of technological development as "a huge mass in rapid motion adjusting to which society with its numerous institutions and organizations has an almost impossible task."[56] Likewise an approach which views technological development as the result of institutional dynamics such as strategies of industrial, political, academic, and military organizations and their interactions, could overcome what seems a deadlock debate over 'knowledge-push' and 'demand-pull' models of change. Not only can examples be found to support both but the sociological approach can easily account for the co-existence of both mechanisms.

The integrating analytical framework is that of a systems theory of social processes as communication in which the production of technical knowledge is related to different types of institutionalized orientations. While Parsonian systems theory treated technology as a borderline structure between the social system and its physical environment, there is no conceivable reason why the explanation of technological change (as well as its social ramifications) should not be an integral part of sociological theory.[57]

A second conclusion concerns the difference between the study of science and the study of technology as knowledge. As I have tried to show the latter cannot be conceived as a self-contained narrowly delineated enterprise. Because of its amorphous nature, its ramifications into diverse institutional areas of society and the heterogeneity of its own institutional bases there is little reason to try to approach technology with the same analytical means as science. At most, these may serve as a heuristic matrix which throws light on characteristic differences and opens research problems. If the concept of 'paradigm', which is ambiguous enough in the study of science, does not apply to technology, it is still fruitful to study persisting concepts, models and solutions in the engineering sciences which guide the perception of engineers, or the role of 'theoretical engineering' as a source of 'autonomous' development of technology, or the importance of 'styles' in technical development which resist changes in the non-technical orientations. A wide field of study is, of course, the micro-analysis of the inventive activity in order to clarify the operation of non-technical orientations as they give direction to the development of technical ideas and solutions. The concept of 'community' does not seem directly applicable either, but in looking for the social structures which represent the institutionalized cognitive orientation complexes one must focus on typical patterns of communication between the educational establishments for engineering and industrial and government laboratories and their respective 'policies' of technical development, as these organizations are the agents of knowledge production.

Finally, technology does not get into 'crisis' by elaboration of concepts. On the contrary, development as the analogous process is invariably the adaptation of the laboratory model to the realistic conditions under which it will eventually have to operate. The adaptation may prove impossible or the conditions may change in the meantime and supersede the original rationale of the whole technique. The study of such failures and of the 'death' of technologies will direct attention to the interdependency of non-technical orientations and their differential impact on the development of technologies when they change. The common focus on 'success stories' usually excludes these aspects.

The structure of technological change is so markedly different from that of scientific change that a transfer of approaches from the study of science to the study of technology seems to have very little promise. However, there are commonalities between science and technology which point to a sociology of scientific and technical knowledge as a common analytical framework. When looking around for ancestors in the field it comes to mind that sociology must have had a particularly bad year in 1935: it missed not only Ludwik Fleck but S.C. Gilfillan, too.

University of Bielefeld

NOTES AND REFERENCES

*I want to thank Edwin T. Layton and the participants of the Workshop 'Models of Scientific and Technological Change', University of Pittsurgh, April 10th - April 12th, 1981, as well as my colleagues at the Science Studies Unit, University of Bielefeld for many valuable suggestions in response to an earlier draft of this paper, being well aware that it still falls short of meeting each of their criticisms.

1. See the pertinent analysis of H. Linde, *Sachdominanz in Sozialstrukturen*, (Tübingen, 1972). Cf. Talcott Parsons and E.A. Shils, eds., *Toward a General Theory of Action*, (New York and Evanston, 1962).
2. Aside from Gilfillan's early attempt at a 'sociology of invention', Thomas P. Hughes may serve as an example of a historian of technology trying to incorporate a systems approach' (though non-sociological) into the history of technology. Cf. S.C. Gilfillan, *The Sociology of Invention*, first ed., 1935. (Cambridge, Mass.: MIT Press, 1970).
3. New attempts to base the sociological explanation of scientific development entirely on an ethnomethdological approach are unconvincing, even though they provide a number of new insights. See Karin Knorr, *The Manufacture of Knowledge. An Essay on the Constructivist and Contextual Nature of Science* (Oxford and Philadelphia: Pergamon Press,, 1980).
4. See Peter Weingart, 'On a Sociological Theory of Scientific Change,' in Richard Whitley, ed., *Social Processes of Scientific Development* (London and Boston: Routledge and Kegan Paul, 1974), 45-68. The interpretation of Kuhn's 'theory' which I have given there accommodates the valid criticisms of his basic categories while retaining the basic tenets.

5. Henry K. Skolimowski, 'The Structure of Thinking in Technology,' in Carl Mitcham and Robert Mackey, eds., *Philosophy and Technology: Readings in the Philosophical Problems of Technology* (New York: Free Press, 1972), 44. This view is doubtful from a constructivist viewpoint with respect to 'a given reality'. That would have a bearing on the concept of science and is of no concern to the argument here. Problems of differentiating between science and technology arise chiefly from phenomena such as 'science based industries' and a 'scientification of technology'. As will be shown below they do not interfere with our approach.
6. This distinction may serve a heuristic pupose. In the case of applied research in industrial and government laboratories it leads to conceptual problems. In a similar vein Polanyi uses the notion of 'operational principles'. They are 'rules' determining action but for Polanyi they are specific to technology and thus much narrower. Michael Polanyi, *Personal Knowledge*, 3rd ed. (London, 1969), 176.
7. ... "*ein Verzeichnis aller der verschiedenen Absichten, welche die Handwerker und Künstler bey ihren verschiedenen Arbeiten haben, und daneben ein VerzeichniB aller der Mittel, durch welche sie jede derselben zu erreichen wissen*. J. Beckmann, *Entwurf der allgemeinen Technologie*, (Göttingen, 1806), 5. (my italics)
8. Cited by Lynwood Bryant, 'The Development of the Diesel Engine,' *Technology and Culture*, 17, (1976), 444.
9. Nathan Rosenberg, 'The Direction of Technological Change: Inducement Mechanisms and Focusing Devices,' in Nathan Rosenberg, *Perspectives on Technology*, (Cambridge: Cambridge University Press, 1976), 110.
10. See F. Redtenbacher, *Prinzipien der Mechanik und des Maschinenbaus* (Mannheim, 1852), 257.
11. A. Bronner, cited by S. Krämer-Friedrich, "*Vergesellschaftung der Natur und Natur der Gesellschaft - Versuch einer gesellschaftstheoretischen Bestimmung der Technik*." Unpublished Dissertation, University of Marburg, 1979, 271-72. See also for the discussion of the literature in construction-science with respect to the impact of economic criteria.
12. Ibid., 274.

13. One case among many where this is explicitly stated is the story of 'Project Whirlwind'. See Thomas M. Smith, 'Project Whirlwind: An Unorthodox Development Project,' *Technology and Culture* 17 (1976), 461.
14. See Fredrich Klemm, *'Technische Entwurfe in der Epoche des Manierismus, besonders in der Zeit zwischen 1560 und 1620,'* F. Klemm, ed., *Zur Kulturgeschichte der Technik,* (Deutsches Museum, München, 1979), 149-164. It is also held that they represented not just toys but models, i.e. tangible theories. Thus, clocks were religiously motivated models.
15. Cyril S. Smith, 'Art, Technology and Science: Notes on their Historical Interaction,' *Technology and Culture* 11 (1970), 493-549.
16. Cf. L. Suhling, *'Deutsche Baukunst. Technologie und Ideologie im Industriebau des "Dritten Reiches",'* in H. Mehrtens and S. Richter, eds., *Naturwissenschaft, Technik und NS-Ideologie,* (Frankfurt, 1980), 243-281.
17. See T. Shinn, *L'École Polytechnique 1794-1914* (Paris, 1980), 53.
18. Cited in K.H. Manegold, *Universität, Technische Hochschule und Industrie* (Berlin, 1970), 153.
19. See Edwin T. Layton Jr., 'Millwrights and Engineers: Science, Social Roles, and the Evolution of the Turbine in America,' in W. Krohn, E.T. Layton Jr. and P. Weingart, eds., *The Dynamics of Science and Technology* (Dordrecht, Holland: Reidel, 1978), 64 ff.
20. See Edward W. Constant II, *The Origins of the Turbojet Revolution* (Baltimore: Johns Hopkins University Press, 1980).
21. See Constant, 'Communities and Hierarchies: Structure in the Practice of Science and Technology,' this volume.
22. Ron Johnston, 'The Internal Structure of Technology,' in Paul Halmos, ed., *The Sociology of Science. The Sociological Review Monograph 18,* (University of Keele, 1972).
23. The example was given to me by Ed Layton in a private communication.
24. See K. Knorr, *Manufacture of Knowledge.*
25. Cf. for Germany, P. Lundgreen, *Techniker in PreuBen während der frühen Industrialisierung* (Berlin 1975).

26. Gilfillan, *Sociology of Invention*.
27. Richard Nelson, M.J. Peck, and E.D. Kalacheck, *Technology, Economic Growth and Public Policy* (Washington, D.C., 1967), 58.
28. Statutes of the VDI, no. 1, cited in N.G. Hortleder, *Das Gesellschaftsbild des Ingenieurs. Zum politischen Verhalten der technischen Intelligenz in Deutschland* (Frankfurt, 1970), 19.
29. Hortleder, *Gesellschaftsbild des Ingenieurs*, 22 ff. and 49 ff. Cf. for the very similar American case, David Noble, *America by Design* (New York: Knopf, 1977), chapter 3.
30. The social structure of biotechnology in Germany is said to resemble a 'club', oriented to the common subject matter but representing common interests vis a vis the universities, funding agencies and government. Reputation in that particular field is gained through the realization of technical processes and new products in industry. See K. Buchholz, *'Die gezielte Förderung und Entwicklung der Biotechnologie,'* in W. van den Daele, W. Krohn, and P. Weingart, eds., *Geplante Forschung* (Frankfurt, 1979), 107. Noble makes the identical observation for the electrical, chemical and mechanical engineers in the United States, Noble, *America by Design* 37 f. He states that the tendency toward professionalization which began in the middle of the nineteenth century "increased the autonomy of scientists, however, it had the opposite effect upon engineers, tying them to the large corporation."
31. Edwin T. Layton Jr., *The Revolt of the Engineers*, (Cleveland: Case Western Reserve University Press, 1971), 19.
32. Constant, *Origins of the Turbojet Revolution*, 9.
33. Ibid., 11.
34. On the concept of 'hybrid communities' cf. van den Daele *et al.*, *Geplante Forschung*, chapter 1.
35. Lynwood Bryant, 'Rudolf Diesel and his Rational Engine,' in *Scientific Technology and Social Change. Readings from Scientific American* (San Francisco, 1974), 116, 118.
36. Lynwood Bryant, 'The Origin of the Automobile Engine,' in *Scientific Technology and Social Change*, *op. cit.*, 109. Otto himself was an amateur inventor.
37. Bryant, 'Development of the Diesel Engine,' 435.

38. This is elucidated very convincingly by stressing the 'development' aspect of invention as Bryant does in his case study on Diesel, and also the contributors Volume 17 of *Technology and Culture* devoted. cf. footnote 8 above. From another angle this is supported by Hart's analysis of 171 inventors mentioned in Kaempffert's *Popular History of American Invention*. He found that "about 3/5 were old hands in the work affected by their inventions, about 1/5 were professional inventors, and another fifth were amateurs or outsiders to the industry". Cited in Gilfillan, *Sociology of Invention*, 87.
39. See R.R. Nelson, 'The Link between Science and Invention: The Case of the Transistor,' *The Rate and Direction of Inventive Activity: Economic and Social Factors*, (National Bureau of Economic Research Publications in Reprint: Princeton, 1962), 559-60.
40. W.O. Baker, 'Science and Technology,' *Daedalus* 109 (Winter 1980), 85.
41. The operation of political orientation complexes, above all, may explain why citation patterns of technology, in contrast to those of science, are nation-specific, and with them, of course, the observable differences of the same technologies.
42. Constant, 'Communities and Hierarchies,' 5. Other examples are: Bessemer steel and subsequent innovations came from outside the iron producing industries, railroads from outside canal and horse transport, the electric light from outside organizations concerned with illumination. Ed Layton, in a private communication.
43. Thomas P. Hughes, 'Inventors: The Problems They Choose, the Ideas They Have and The Inventions They Make,' in Patrick Kelly *et al.*, eds., *Technological Innovation: A. Critical Review of Current Knowledge*, (San Francisco: San Francisco Press, 1978), 166-182. Hughes draws on A.P. Usher, *A History of Mechanical Inventions* (London, 1954), chapter 4; S.G. Gilfillan, *Sociology of Invention*, and Thomas P. Hughes, *Elmer Sperry: Inventor and Engineer*, (Baltimore: Johns Hopkins University Press, 1971).
44. Usher, *History of Mechanical Inventions*, 71.
45. Hughes, 'Inventors,' 169.
46. Gilfillan, *Sociology of Invention*, 7.
47. Hughes, 'Inventors', 172, 173.

48. This concurs in part with Johnston, 'Internal Structure', 124.
49. Thomas P. Hughes, *The Electrification of America: The System Builders*, unpubl. ms., 25.
50. Nathan Rosenberg, 'Factors Affecting The Diffusion of Technology,' in Rosenberg, *Perspectives on Technology*, 206, 231.
51. Constant, *Origins of the Turbojet Revolution*.
52. *Ibid.*, 15 ff.
53. Gilfillan, *Sociology of Invention*, 45.
54. See Nathan Rosenberg, 'Direction of Technological Change,' 121; also on the effect of materials shortages on the inducement of new technologies, Nathan Rosenberg, 'Innovative Responses to Materials Shortages,' in: Nathan Rosenberg, *Perspectives on Technology*, 253. See these for examples.
55. N. Rosenberg makes a similar suggestion with respect to economic studies of technological change in 'Direction of Technological Change', and Hughes' thrust is very similar when he says that "any analysis be one that is shaped ... by a dynamic systems model" in *Technological Innovation*, 180.
56. W.F. Ogburn, *On Culture and Social Change* (Chicago and London, 1964), 85. For analysis and critique of various 'theories' of autonomous technological development see Langdon Winner, *Autonomous Technology* (Cambridge, Mass.: MIT Press, 1977), esp. chapter 2.
57. Talcott Parsons, *Societies. Evolutionary and Comparative Perspectives* (Englewood Cliffs, New Jersey, 1966), 16.

AUTHOR INDEX

SOCIOLOGY OF THE SCIENCES
MONOGRAPHS

Already published in this series:

Marc de Mey, *The Cognitive Paradigm*
1982, xx + 314 pp., ISBN 90-277-1382-0

Tom Jagtenberg, *The Social Construction of Science*
1983, xviii + 237 pp., ISBN 90-277-1498-3

Norman Stockman, *Antipositivist Theories of the Sciences*
1983, x + 284 pp., ISBN 90-277-1567-X